AF690166

Journal of Applied Logics - IfCoLog Journal of Logics and their Applications

Volume 13, Number 1

January 2026

Disclaimer

Statements of fact and opinion in the articles in Journal of Applied Logics - IfCoLog Journal of Logics and their Applications (JALs-FLAP) are those of the respective authors and contributors and not of the JALs-FLAP. Neither College Publications nor the JALs-FLAP make any representation, express or implied, in respect of the accuracy of the material in this journal and cannot accept any legal responsibility or liability for any errors or omissions that may be made. The reader should make his/her own evaluation as to the appropriateness or otherwise of any experimental technique described.

ISBN 978-1-84890-501-6
ISSN (E) 2631-9829
ISSN (P) 2631-9810

College Publications
Scientific Director: Dov Gabbay
Managing Director: Jane Spurr

http://www.collegepublications.co.uk

Editorial Board

Scope and Submissions

This journal considers submission in all areas of pure and applied logic, including:

- pure logical systems
- proof theory
- constructive logic
- categorical logic
- modal and temporal logic
- model theory
- recursion theory
- type theory
- nominal theory
- nonclassical logics
- nonmonotonic logic
- numerical and uncertainty reasoning
- logic and AI
- foundations of logic programming
- belief change/revision
- systems of knowledge and belief
- logics and semantics of programming
- specification and verification
- agent theory
- databases
- dynamic logic
- quantum logic
- algebraic logic
- logic and cognition
- probabilistic logic
- logic and networks
- neuro-logical systems
- complexity
- argumentation theory
- logic and computation
- logic and language
- logic engineering
- knowledge-based systems
- automated reasoning
- knowledge representation
- logic in hardware and VLSI
- natural language
- concurrent computation
- planning

This journal will also consider papers on the application of logic in other subject areas: philosophy, cognitive science, physics etc. provided they have some formal content.

Submissions should be sent to Jane Spurr (jane@janespurr.net) as a pdf file, preferably compiled in LaTeX using the IFCoLog class file.

Contents

Editorial

Paolo Fosci
University of Bergamo, Italy
paolo.fosci@unibg.it

Sérgio Ramos
Polytechnic Institute of Porto, Portugal
scr@isep.ipp.pt

Paweł B. Myszkowski
Wrocław University of Science and Technology, Poland
pawel.myszkowski@pwr.edu.pl

Héctor Quintián
University of A Coruña, Spain
hector.quintian@udc.es

Special Issue FDOM 2024

FDOM 2024 is a significant forum for presenting the development and applications of innovative techniques in closely related areas. The exchange of ideas between scientists and technicians from both academic and business sectors is essential to facilitate the development of systems that meet the demands of today's society. Technology transfer in this field remains a challenge, so such contributions are mainly considered in this symposium. The FDOM Special Session features discussions and publications on developing innovative techniques for complex problems with special focus on using data-driven models and artificial intelligence to identify system failures, improve operational efficiency, and make informed decisions, especially in industrial and complex systems, by covering techniques like machine learning, time-series analysis, and statistical process control to build models that can predict faults, optimize performance, and aid in fault diagnosis, which are crucial for safety, reliability, and cost reduction.

This Special Issue includes nine papers selected from extended contributions presented at the Special Session Fault Detection, Optimization, and Modeling for Decision Making (FDOM) under the framework of the 15th International Symposium on Ambient Intelligence 2024 (ISAMI 2024), held in Salamanca, Spain, March 15th, 2024.

The contributions are organised as follows.

The first contribution, by Michelena et al., addresses the challenge of early estrus detection in dairy cows through behavioral data collected via smart collars. The study evaluates various supervised classification techniques combined with feature engineering, using data from twelve cows on a Galician intensive farm. Individualized models were developed for each animal, and different configurations of models and input data were tested. Results showed excellent performance, with some configurations achieving F1-Scores of up to 1. The findings underscore the importance of personalized modeling and incorporating temporal behavioral context to enhance detection accuracy.

In the second contribution, by Aveleira-Mata et al., the focus is on enhancing security in Zigbee-based IoT networks by detecting denial-of-service (DoS) attacks using supervised machine learning techniques. The authors generated a specialized dataset from a simulated Zigbee environment, capturing both normal and malicious traffic. Six classification algorithms were evaluated—Logistic Regression, K-Nearest Neighbors (KNN), Support Vector Machines (SVM), Decision Trees, Random Forests, and Artificial Neural Networks (ANN). Among these, KNN, Decision Trees, and Random Forests demonstrated superior accuracy and efficiency. The study emphasizes the value of tailored datasets and machine learning approaches for safeguarding IoT systems against evolving cyber threats.

The third contribution, by Mohammed et al., focuses on optimizing machine learning processes where evaluating each solution is computationally expensive or time-consuming. The study applies Bayesian optimization to fine-tune the hyperparameters of Convolutional Neural Networks (CNNs), aiming to enhance performance while conserving resources. It first examines the effectiveness of the Stochastic Gradient Descent with Momentum (SGDM) optimizer and highlights the critical role of the momentum hyperparameter. Bayesian optimization is then used to model CNN generalization performance as a Gaussian process, updating the surrogate model based on limited evaluations. The results demonstrate that this approach yields promising outcomes for hyperparameter tuning under resource constraints.

The following paper, by Pérez et al., investigates the use of neural networks to improve adaptive optics systems in ground-based telescopes, which are affected by atmospheric turbulence that blurs images. The study employs a 2D-LSTM network to predict wavefront slopes over time, aiming to forecast turbulence while filtering out noise from measurement instruments. Using a dataset of 40,000 clean and 40,000 noisy instances simulated with Soapy, the experiments demonstrate that the neural models can effectively learn noise patterns and produce outputs that closely resemble actual turbulence conditions, enhancing the accuracy of atmospheric correction.

The fifth contribution, by Arcano-Bea et al., explores the use of computer vision technologies to enhance component identification and traceability in the shipbuilding industry. The authors propose a dual approach that integrates 3D point cloud analysis with 2D deep learning-based instance segmentation. Using a depth camera, they acquired both 3D and high-resolution 2D data. Surface and edge matching techniques were applied for the 3D analysis, while advanced models such as YOLOv1 and Mask R-CNN were fine-tuned for the 2D segmentation. The combined methodology significantly improves the reliability of subassembly detection, contributing to better traceability, reduced errors, and more efficient production workflows in shipbuilding.

The next contribution, by Oliveira et al., addresses the growing concern of air pollution, particularly Nitrogen Dioxide (NO_2), which poses serious health and environmental risks. The study focuses on forecasting NO_2 concentrations in Porto using five Deep Learning (DL) models: Multi-Layer Perceptron (MLP), LSTNet, Temporal Convolutional Networks (TCN), Transformers, and a hybrid Transformer-LSTM architecture. These models were designed, tuned, and evaluated to predict pollution levels up to two days in advance. Among them, the MLP model achieved the best performance, with a Root Mean Square Error (RMSE) of 7.65 $\mu g/m^3$. The findings highlight the effectiveness of DL techniques in environmental monitoring and their potential to support informed decision-making and pollution mitigation strategies.

The seventh contribution, by Gil-Arroyo et al., proposes the use of Deep Learning models to automate fabric quality control in the textile industry, where inspections are still largely dependent on human visual assessment. The study evaluates Convolutional Neural Networks using real images from Batavia and Sarga weave production lines. Experimental results identify DenseNet121 as the most effective model for Batavia weaves, offering balanced performance across key metrics, while InceptionV3 excels in detecting defects in Sarga weaves, particularly in terms of F1-score and AU-ROC. These findings demonstrate the potential of DL models to improve the accuracy and efficiency of textile quality assurance processes.

The subsequent paper, by Loukanova, presents an extended version of the Type-Theory of Acyclic Recursion and Acyclic Algorithms (TTAR/TTAA), introducing a new chain-reduction rule to its reduction calculus. TTAA enables stepwise, mutually recursive computations with results stored in memory slots for reuse. The extended calculus eliminates redundant chain-like assignments of functional values, improving computational efficiency. This refined framework is particularly relevant for the algorithmic semantics of formal and natural languages, including applications in programming languages and compiler components that transform recursive programs into iterative ones.

The last paper, by Torres et al., provides an overview of how Generative Artificial Intelligence (GenAI) techniques can be used to produce content in a controlled and reliable manner, particularly for applications requiring high accuracy. While GenAI has transformed various domains by generating text, images, video, and audio, it is prone to producing false or inconsistent outputs—commonly referred to as "hallucinations." To address this, the authors propose a methodology that enables users to guide GenAI systems using controlled, user-defined sources. The approach is illustrated through an educational use case: the automatic generation of multiple-choice exams. This work highlights the importance of controlled content generation in sensitive contexts and demonstrates a practical solution for ensuring reliability.

The guest editors thank Professor Dov Gabbay (Founding Editor-in-Chief of *Journal of Applied Logics - IfCoLog Journal*) for allowing the opportunity to edit this special issue. We would also like to thank the referees who have critically evaluated the papers within a short time. Finally, we hope the reader will share our joy and find this special issue very useful.

Received December 2025

Novel real-time heat detection system for dairy cows based on supervised techniques and smart collars

Álvaro Michelena*
University of A Coruña, CTC, CITIC, Department of Industrial Engineering
alvaro.michelena@udc.es

Óscar Fontenla-Romero
University of A Coruña, LIDIA, CITIC, Faculty of Computer Science
oscar.fontenla@udc.es

José Luis Calvo-Rolle
University of A Coruña, CTC, CITIC, Department of Industrial Engineering
jose.rolle@udc.es

Abstract

In precision animal husbandry, behavioral monitoring collars have become widely used tools for individualized and real-time herd management. These devices allow for the collection of individual animal activity data to be analyzed to identify behavioral patterns or specific events of interest. One of the most relevant events that can be detected by analyzing behavioral changes is estrus.

In this context, the present paper analyzes and compares the performance of several supervised classification techniques combined with feature engineering for early estrus detection in dairy cows from intensive farms using behavioral data obtained from a commercial smart collar. For this purpose, twelve dairy cows were monitored in a Galician intensive farm, and customized models were developed for each animal. Several configurations of both the models and the input data used were evaluated.

The results obtained were excellent, reaching, in some cases, F1-Score values of up to 1 in specific configurations and techniques. In addition, this research demonstrates a high variability in model performance between cows, highlighting the need to develop individualized animal models. It is also concluded that to obtain good results, it is essential to provide the model with a temporal context that includes the animal's previous behavior.

*Álvaro Michelena's research was supported by the Spanish Ministry of Universities (https://www.universidades.gob.es/), under the "Formación de Profesorado Universitario" grant with reference FPU21/00932.

1 Introduction

The intensive farms predominate over the rest of the management systems in the current livestock sector. These farms are characterized by housing many animals in reduced areas. This intensive model is widely implemented due to maximized production and reduced costs, offering higher profitability than extensive systems.

However, this type of management approach poses significant challenges. The high density of animals confined in fairly small spaces can compromise their welfare, causing stress or disease problems, among others. Because of this, the abusive use of antibiotics to minimize the risk of infection and prevent the spread of disease is widespread [1]. This practice also brings challenges and is currently being questioned and criticized. In addition, the environmental impact of these farms is considerable, generating significant greenhouse gas emissions, such as methane, which contribute to climate change [2].

Faced with these issues, advanced technologies are being implemented to reduce environmental impact and promote better animal welfare through more ethical and sustainable practices. In this sense, one of the key strategies is adopting innovative technologies, such as sensors, monitoring devices, data analysis, and artificial intelligence tools, which allow obtaining real-time information on the animal's health, behavior, and productivity. This approach, known as Precision Livestock Farming (PLF), has acquired relevance in modern livestock farm management [3].

Within the PLF, intelligent monitoring collars have proven to be especially useful. These devices, equipped with different sensors, among which accelerometers and gyroscopes stand out, are capable of detecting the daily activities of each animal in real-time, and their use has become popular due to their effectiveness, ease of use, and affordable cost compared to other technological tools [4].

Thanks to these smart collars, it is possible to monitor individual animal behavior in real-time and detect patterns that deviate from normal behavior. These unusual behaviors can be associated with specific events, such as diseases or reproductive processes, such as parturition or estrus [5].

In this context, one of the main advances has been analyzing animal behavior to detect health conditions or reproductive events. Recent research indicates that sick animals tend to reduce or modify their activities in order to conserve energy, which can result in time changes in key activities [6]. On the other hand, it has also been observed that cows significantly reduce rumination time on calving days, increasing rest time compared to the gestation period [7].

Monitoring rumination time combined with other activities allows early identification of metabolic problems, digestive problems, and severe cases of mastitis and metritis, even days before clinical symptoms manifest themselves [8]. Other aspects of behavior, such as resting time, locomotion, feeding, and social interaction, also provide valuable indicators of animal welfare and health [9, 10, 11].

Therefore, the analysis of animal behavior allows the identification of patterns associated with different events of interest, which makes the detection and analysis of behavioral changes an essential aspect of modern animal husbandry [12].

1.1 Estrus in cattle

In the cattle industry, one of the most essential reproductive events is estrus and its absence, known as anestrus. Estrus is the period within a cow's reproductive cycle during which she is receptive to reproduction [13].

Cows have estrus cycles throughout the year, although periods of fertility are limited. Estrus detection and prediction are, therefore, critical to optimizing productivity on a dairy farm and maximizing profitability. The estrous cycle, which encompasses the physiological events between one estrus and the next, lasts 21 days. However, these can vary between 18 and 24 days in cattle. Under normal conditions, cows reach puberty between 9 and 15 months [14].

In cows, the estrus period is brief, lasting an estimated 12 to 18 hours. During this time, cows have signs and behavioral changes that reflect their predisposition for mating [14, 13]. These include increased activity (both in steps and reduced resting time) and decreased feed intake and rumination time [15]. In addition to these behavioral changes, the cow undergoes alterations in estrogen and progesterone hormone levels, temperature variations, and an increase in vaginal mucus production [16].

These hormonal changes influence the cow's behavior during estrus. They are also valuable indicators for determining the right time for artificial insemination or natural mating, optimizing the likelihood of reproductive success. Accurate estrus monitoring is crucial in intensive farms, where artificial insemination is standard. Accurate estrus detection is therefore crucial to improve the success rate of these procedures.

In addition, monitoring estrus allows the identification of anestrus, a condition commonly occurring during the postpartum period and is one of the leading causes of infertility in dairy cattle. Although a brief estrus-free period after calving is expected, a prolonged anestrus, more than 60 days after calving, can negatively affect the animal, resulting in significant production losses [17, 18].

Traditionally, detection of estrus was done by visual observation. However, factors such as reduced estrus time and the increase in the number of animals per farm have made visual detection more complex, prompting the development of automated systems to detect estrus more accurately [19, 20].

These automated systems rely on various sensors to capture cows' different signs and behavioral changes. The most commonly used devices are collars with accelerometers, pedometers, and ear tags. In [21], the HeaTime® activity monitoring system detected 94 of 111 estrus in free-stall Holstein heifers using an empirically adjusted threshold that generated real-time high activity alerts.

Similarly, a commercial device based on 3D accelerometers designed to be placed on ear tags was evaluated in [22]. The collected movement data were processed with different algorithms to identify behavioral patterns related to estrus. Accelerometers have also been combined with other sensors in more robust hybrid systems to improve detection sensitivity and reduced false alarms [23].

In addition, the development of artificial vision has allowed the application of new image-based methods. Combining video cameras with advanced computer vision techniques has shown potential for heat detection. Likewise, thermography and other temperature sensors, either at the base of the tail or intravaginal sensors, have also been investigated to detect estrus by monitoring parameters such as temperature and conductivity of the vaginal mucosa [24, 25, 16].

Considering the importance of estrus in the profitability and efficiency of livestock farms, this research work focuses on analyzing and comparing the performance of five supervised classification techniques for early estrus detection using behavioral data collected through a smart collar.

This paper is structured as follows. After the introduction, Section 2 provides a detailed description of the monitoring device used and all relevant information about the dataset used in this research work. Section 3 briefly describes each of the five supervised classification methods evaluated. Section 4 describes in detail the steps and procedures followed in the execution of the experiments. The results obtained from these experiments are presented and analyzed in Section 5. Finally, Section 6 provides conclusions and suggests possible lines of future work.

2 Case of Study

This section presents a detailed description of the smart monitoring collar used, highlighting its main features and practical functionalities. On the other hand, all relevant information about the dataset used in the development of this research is also included.

2.1 Livestock smart collar

Among the devices used to monitor livestock behavior, one of the most common are smart collars. Currently, there are many companies dedicated to the development of this kind of devices. In this research work we had the support of the Galician startup Innogando, which has designed a monitoring collar, called RUMI, specially designed for monitoring cattle [26].

The RUMI collar, Figure 1, is an intelligent device designed to monitor the behavior of cows in real time. Equipped with GPS technology and high-precision sensors, such as accelerometers, the collar allows recording several key activities: resting, rumination, grazing, walking and a spontaneous and random behavior. In addition, the device counts the animal's steps, providing a complete overview of its daily movements and activities.

Figure 1: *Innogando* Smart Collar

The collar is designed to be placed on the left side of the cow's neck, secured with a strong nylon strap that prevents skin irritation and allows full movement freedom. To ensure its position, a counterweight is attached to the underside of the collar. With a light weight and robust construction, the RUMI can withstand extreme conditions, thanks to its IP67 protection against rain, dust and UV rays.

One of the highlights of this device is its automatic recharging capability via an integrated solar panel, which ensures long life without the need to change or recharge the batteries. Even in low light conditions, it can work for up to eight months due to its energy efficiency.

For data transmission, the collar uses the LoRaWAN low-power communication protocol, sending the information to a central server. In this way, the data can be accessed in real time from a mobile application or laptop, facilitating farm management and animal tracking.

Table 1 provides the main specifications of this device.

RUMI main features	
Dimensions	130 x 90 x 37mm
Device weight	350 g
Counterweight	460 g
Total weight	810 g
Operating temperature	-20ºC to 50ºC
Casing material	ABS
IP protection	IP67
Fixing method	Belt
Recharge mode	Solar panel
Battery life (without charge)	240 days
Communication protocol	LoRaWAN

Table 1: RUMI main features

2.2 Dataset

For this research, RUMI Smart Collar was attached to a total of 12 cows on an intensive dairy farm located in Galicia, Spain. The farm operates in an intensive management system, which implies that the animals remain in an indoor barn with no access to outdoor pasture.

Monitoring was conducted from March to September 2024. The 12 cows were carefully selected, ensuring that none were pregnant at the start of monitoring and that they represented different ages, thus guaranteeing greater variability and quality in the dataset. All cows monitored belonged to the Holstein Friesian breed, the most common breed in the dairy industry due to its high milk production.

To ensure the representativeness of the dataset, standard dairy farm procedures were maintained during monitoring, without introducing additional interventions or external practices. In this way, data were obtained that accurately reflect the usual behavior of cows on this kind of farm.

The monitoring collar recorded the behavior of each cow on an hourly, providing a total of eight different variables, listed below:

- **Collar ID**: Unique identifier of each collar, which allows assigning each record to the corresponding animal.
- **Time stamp**: Date and time linked with each record.
- **Resting time**: Minutes the cow spent resting in a given hour.

- **Ruminating time**: Minutes the cow spent ruminating in a given hour.
- **Walking time**: Minutes the cow spent moving within an hour.
- **Grazing time**: Minutes the cow was engaged in grazing activity within an hour.
- **Time remaining**: Minutes the cow spent in activities other than the above. Within the dataset, this variable is referred to as *Eat and Play*, and is related to sudden and random movements.
- **Number of steps**: Total number of steps taken by the cow in one hour.

It is important to note that, due to the intensive farm management, the animals do not go out to graze, but the *Grazing time* variable is important to consider since the collar technology often detects this behavior when the animal is eating food provided by the farmer such as feed or dry grass.

Since the aim of this investigation is the evaluation of supervised techniques for heat detection, the dataset has been labeled. Innogando staff, together with farm workers and under the supervision of veterinarians, were in charge of this labeling process, assigning each record a binary variable indicating whether the behavior corresponds to an estrus (1) or not (0).

During the monitoring period, some of the cows became pregnant, so when pregnancy was detected, the recording of their behavior was interrupted to avoid introducing noise in the dataset. Thus, each animal has a different number of records and estrus events monitored.

Finally, Table 2 shows the total number of records per cow, the number of monitored estrus events and the total number of records corresponding to estrus.

As can be seen in Table 2, and as expected, the dataset suffers from a large imbalance of data between normal behavior, the majority class, and estrus records, the minority class. This aspect is essential to take into account when training and evaluating supervised classifiers.

3 Supervised classification methods

In this research work, the performance of five supervised classification techniques, widely used in several areas, has been evaluated and compared. This section presents a brief description of each of these techniques.

Cow ID	Estrus monitored	Total hours monitored	Estrus hours monitored
1	4	1632	47
2	3	1152	43
3	3	1080	49
4	3	1152	38
5	5	2352	50
6	3	1464	54
7	3	1128	34
8	5	2496	89
9	4	1656	54
10	4	2088	72
11	10	4992	181
12	8	3360	133

Table 2: Samples per animal

3.1 Logistic Regression

Logistic regression is a method of statistical analysis applied in the resolution of classification problems [27]. This approach uses logistic or sigmoid functions, which generate an output value restricted between 0 and 1, allowing it to be interpreted as a probability.

It is relevant to note that logistic or sigmoid functions are used because they are derivable at any point in their domain, unlike other functions, such as the step function, which lack this property.

Since the result of the logistic or sigmoid function is expressed as a probability, to carry out the final classification it is necessary to establish probability thresholds, which define the decision limits.

The training of this model is performed by means of the gradient descent algorithm, used to minimize the error function (cross-entropy), defined as the negative natural logarithm of the probability [28].

3.2 Decision Trees

Decision trees are one of the simplest and most common supervised machine learning techniques. This method is based on building a model with a hierarchical tree structure that includes a root node, branches, decision nodes and output nodes, also known as leaves [29].

The process starts at the root node, where one dataset variable is evaluated. Depending on the value of this variable, one of the branches is selected to continue the analysis at a decision node. This procedure is repeated until the data reaches a leaf node, where a classification is assigned based on the value of that node. The decision tree segments the data according to the values of its variables, so it is crucial to determine the optimal cut-off points. To do this, the method computes metrics such as entropy or the Gini index to measure the impurity of each variable. With these values, the information gain is evaluated by comparing the impurity of the data before and after performing the splits at the nodes. Thanks to its hierarchical structure, the tree is built from the root to the leaves, using first the variables that maximize the information gain [30].

The models generated by the decision tree algorithm are also intuitive and easy to interpret. However, this technique may have limitations in complex problems, as it tends to produce large and detailed trees that often lead to overfitting.

3.3 Random Forest

The Random Forest (RF) method is a supervised machine learning technique widely used for classification tasks and also useful for regression problems. Its approach consists of applying several decision trees, where its final classification class corresponds to the most frequent class among the results of the individual trees [31].

Each decision tree is trained using a different training data samples. The Bootstrap method, also known as Bagging, is used to create these data subsets. This ensemble technique selects random subsets of the whole dataset. It is important to note that this technique allows the same sample to appear in multiple subsets [32].

3.4 Support Vector Machines

Support vector machines, commonly known by their acronym SVM, are a supervised learning technique developed by Cortes and Vapnik to solve both classification and regression problems [33]. In classification tasks, this method focuses on finding the best hyperplane that separates the data by maximizing the distance (or margin) between samples in each class.

When each class's samples in the dataset are linearly separable, SVMs fit a hyperplane that maximizes the margin between classes. However, SVMs also work effectively with data that are not linearly separable. In these situations, the method applies nonlinear transformations to the original space, transforming it into a higher-dimensional space where a linear separation between classes is possible. In order to perform these transformations, different types of kernels are used, the most important being polynomial, sigmoidal and Gaussian kernels [34].

Thanks to their ability to adapt to both linearly separable and non separable datasets, SVMs are widely used in many fields [35].

3.5 Artificial Neural Networks

Artificial Neural Networks (ANN) are mathematical models inspired by the functioning of the human brain and its neural connections. These networks are made up of artificial neurons that interconnect, creating a network or matrix structure [36].

Each artificial neuron follows a simple mathematical process to generate an output from the input data. First, each input is given a specific weight, multiplied by that value. Then, the weighted inputs are summed together with an additional term, known as bias, which yields a single value. This resulting value is transformed through an activation function to generate the output of the neuron. Among the activation functions used are linear or sigmoidal, among others [37].

The neurons are organized in layers, which configures the structure of the neural network. After building the model, it is necessary to train it in order to optimize the weights in each layer. The training is carried out using the backpropagation method, which allows recalculating the weights to minimize the prediction or classification error in the network output [38].

In addition, dropout layers are usually included in these networks to improve regularization and avoid overfitting. These layers randomly deactivate some neurons in the network, and their influence is adjusted by a probability value that controls their effect on the model.

4 Stages and experimental setup

This section presents a detailed description of all the steps and processes implemented to carry out the experiments necessary to obtain the final results.

For a better understanding of this section, it has been divided into several subsections describing the main experimental steps performed in this research work.

4.1 Dataset preprocessing and splitting

The first step was to split the dataset by cow, which made it possible to obtain twelve subsets from the initial dataset, one for each of the monitored animals. This dataset division is necessary because this research aims to evaluate the performance of the supervised techniques described in Section 3 by means of individual cow models, following the scheme presented in Figure 2, which shows an example considering only three monitored cows. This is because behavior among different cows can vary considerably, which could affect the efficacy of these techniques if they were used in a global model that included data from all animals.

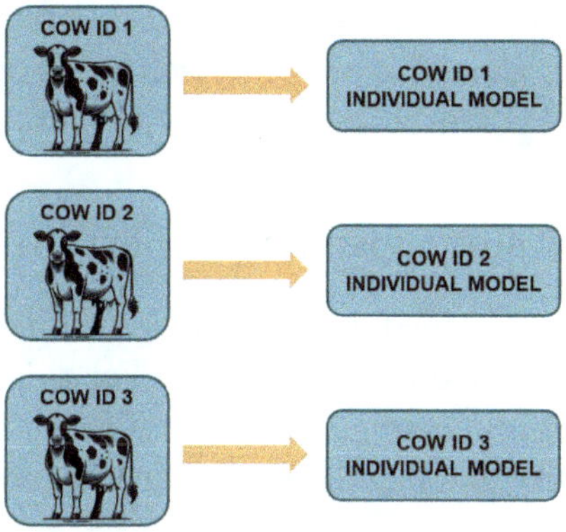

Figure 2: Model topology

Once the data were segregated, each of the twelve subsets was checked for incomplete hourly records. Missing data were completed by replicating the value recorded in the previous hour corresponding to that cow, to preserve temporal continuity and minimize the loss of information.

4.2 Generation of new variables

Once the data had been separated by animal, the next step was to generate new variables from the six initial variables: resting, ruminating, walking, grazing, eating/playing and number of steps. This process, known as feature engineering, improves the model performance by adding additional and potentially more meaningful information.

Six new variables were generated, corresponding to the sum of each activity and the number of steps accumulated in a given interval. This step is essential because estrus behavior is manifested by continuous changes over several hours. Without this data aggregation, classifiers could classify any isolated momentary anomalous behavior as estrus without differentiating between one-time variations and an actual estrus episode. These new variables allow the model to interpret patterns accumulated over hours, providing a broader and more relevant perspective.

To analyze the performance of the supervised techniques, accumulation intervals from 2 to 12 hours prior were evaluated, resulting in multiple data configurations, each representing a different level of temporal analysis.

In addition, an additional numeric variable was incorporated to reflect the time slot of each record. Like many animals, cows exhibit behavioral patterns that vary by time of day, mainly due to circadian rhythms. In intensively managed systems, these time-of-day variations are shown to a greater extent due to the feeding and management routines inherent to these farms. Considering the time of day is relevant for detecting estrus since it usually manifests itself at night or dawn, providing extra information in detecting this kind of reproductive event.

4.3 Configuration of supervised classifiers

Once the different datasets obtained were preprocessed, the selected supervised classification techniques were trained and optimized. For this purpose, the performance of the models was analyzed and compared under different configurations of hyperparameters, looking to identify the best combination that maximizes the performance in early estrus detection.

The hyperparameters and values evaluated for each of the classification algorithms used are detailed below:

- **Logistic Regression**: Different values of the regularization parameter C were explored. The value of this hyperparameter adjusts the model's regularization, avoiding overfitting; the smaller the value of this parameter, the higher the regularization.

  - C: 0.0001, 0.0005, 0.001, 0.005, 0.01, 0.05, 0.1, 0.5 and 1.

- **Support Vector Machine**: Several configurations were tested, adjusting both the type of kernel, which defines the data transformation function applied and the regularization parameter, C. It should be noted that the regularization decreases as the value of C increases.

  - *Kernel*: linear, sigmoidal, polynomial, and radial basis function.
  - C: 0.001, 0.005, 0.01, 0.05, 0.1, 0.5 and 1.

- **Decision Tree**: For this method, different maximum depths were evaluated, which limit the maximum number of levels in the generated tree. The tree's depth directly impacts its ability to capture complex patterns, but excessive depth can lead to overfitting.

   - Maximum tree depth: 2, 3, 4 and [5:5:100]. The self-adjustment of the tree depth was also evaluated.

- **Random Forest**: Configurations were explored by varying the number of trees that make up the forest. The number of trees affects the stability and accuracy of the model, as more trees can reduce bias and variance, improving overall performance.

   - Number of trees: 2, 3, 4 and [5:5:100].

- **Artificial Neural Networks**: Several neural network configurations were developed by setting the number of neurons in each layer, the number of layers, and the dropout value. Adjusting these hyperparameters allows the control of the complexity of the network and reduces the risk of overfitting, thus improving the model's generalization capability.

   - Neurons per layer: 2, 5 and 10.
   - Number of hidden layers: 1, 2 and 3.
   - Dropout: 10% and without dropout.
   - Activation function in hidden layers neurons: *ReLu*
   - Activation function in output layer neuron: *Softmax*

4.4 Training and evaluation of classifiers and selection of best configurations

For each of the datasets obtained and for each model with its different configurations, a training and performance evaluation process was carried out. The aim was to identify the best configurations for each technique and select the best final models. The training process was divided into two main phases:

- First phase: which was carried out to determine the optimal hyperparameter configurations for each technique.

- Second phase: used to evaluate the overall performance of each model using the best configurations obtained in the previous stage.

It is important to note that 60% of the samples from each dataset were selected for the first phase, ensuring that this set included at least one estrus event. Similarly, the remaining 40% (test set) was also verified to contain at least one estrus event, ensuring that both classes were represented in the training and validation samples. The steps carried out in each of the two phases are described below.

4.4.1 First stage. Hyperparameter selection

This first phase aimed to identify the best configurations of each classification technique in each of the different datasets. For this purpose, k-fold cross-validation was used, employing a value of k=3 due to the reduced number of samples in the datasets.

On the other hand, as explained in Section 2.2, the dataset presents a significant imbalance between classes, with fewer records related to estrus events compared to records of normal behavior. This imbalance may lead the model to overfit the majority class, negatively affecting its accuracy in detecting the estrus (minority class).

Two analogous strategies can be implemented to address the imbalance problem [39]. On the one hand, it is possible to increase the number of samples of the minority class by generating additional data following specific patterns (Oversampling). On the other hand, the number of samples from the majority class can be reduced by applying rules that keep the selected data representative (Undersamplig).

In the present research work, given the limited data, the oversampling strategy was chosen, increasing the number of samples of the minority class (estrus) by using the Synthetic Minority Oversampling Technique, commonly known as SMOTE [40]. This technique was applied only to the training folds samples and never to the validation instances.

After applying SMOTE, the training fold samples were normalized using the Z-Score method, which is essential due to the differences in the scale of the dataset variables [41]. For example, the number of steps tends to have significantly higher values than the behavioral time features. This normalization allows the model to work on a uniform scale, preventing it from being influenced by variables with higher values.

With the training data balanced and normalized, the model was trained and evaluated. The F1-Score, defined as the harmonic mean between accuracy and sensitivity, was used as the key evaluation metric. This metric is particularly suitable in anomaly detection contexts, as it combines high sensitivity with high precision, which is essential in unbalanced datasets such as the one used in this research. Equation 1 shows the mathematical description of this metric.

$$\text{F1-score} = \frac{2 \times \text{True Positives}}{2 \times \text{True Positives} + \text{False Positives} + \text{False Negatives}} \tag{1}$$

After training and evaluating all configurations of the same technique, their results were compared to select the configuration with the best performance. For this purpose, the Kruskal-Wallis analysis of variance was applied together with Tukey's comparison method [42]. This analysis allows the detection of significant differences between the models. If differences were detected between the configurations, the one with the highest F1-Score was selected. If no differences were detected, the simplest configuration was chosen regarding computational cost, evaluated by the average training time required.

4.4.2 Second stage. Final performance evaluation

Once the best configurations for each technique were selected, the final models were trained using the whole training dataset (60%). As in the first phase, SMOTE was used to balance the classes and the Z-Score normalization method.

Once the models were trained, the test set (remaining 40%) was used to obtain the final performance results. Finally, the models were compared, and the one with the highest F1-Score was selected.

As a summary, Figure 3 presents schematically the different stages or steps that constituted the execution of the experiments.

5 Analysis of obtained results

This section contains the results obtained in the presented research work. As previously mentioned, multiple configurations have been evaluated, both in the hyperparameter settings of each technique and in the number of hours considered for the generation of the new variables.

Each configuration has been evaluated and analyzed independently for each dataset obtained, considering both different cows and different temporal durations (number of hours considered). This has resulted in a significant volume of results.

To simplify the interpretation of the results, the tables and figures in this section only give the results corresponding to the configurations that achieved the highest F1-Score values. As highlighted in Section 4.4, the metric used is the F1-Score, expressed in a range from 0 to 1. The higher the F1-Score, the higher the classifier's performance, reflecting a better ability of the model to detect estrus events.

Table 3 presents the results corresponding to the three best models for each cow, showing the configurations that achieved the highest F1-Score values.

Table 3: Results of the three best classifiers and configurations for each cow

Cow ID	Hour	Technique	Configuration	F1-score
1	3	SVM	Kernel = linear C = 1.0	0.879

1	11	ANN	Hidden Layers = 2 Neurons per layer = 10 Dropout = 10 %	0.866
1	9	SVM	Kernel = linear C = 1.0	0.864
2	4	DT	Tree levels = 2	0.927
2	3	DT	Tree levels = 2	0.923
2	8	DT	Tree levels = 2	0.917
3	11	ANN	Hidden Layers = 3 Neurons per layer = 10 Dropout = 0 %	0.978
3	11	SVM	Kernel = rbf C = 1.0	0.978
3	10	SVM	Kernel = linear C = 0.5	0.977
4	6	DT	Tree levels = 2	1.000
4	7	DT	Tree levels = 2	1.000
4	12	DT	Tree levels = 2	0.955
5	7	DT	Tree levels = 2	0.882
5	3	DT	Tree levels = 2	0.862
5	3	SVM	Kernel = rbf C = 0.5	0.852
6	5	SVM	Kernel = linear C = 0.5	1.000
6	7	SVM	Kernel = linear C = 0.5	0.981
6	4	SVM	Kernel = linear C = 0.5	0.957
7	8	LR	C = 1.0	0.930
7	12	ANN	Hidden Layers = 3 Neurons per layer = 10 Dropout = 0 %	0.913
7	8	ANN	Hidden Layers = 3 Neurons per layer = 10 Dropout = 0 %	0.905
8	7	RF	Number of trees = 2	0.845
8	11	DT	Tree levels = 2	0.838

8	11	SVM	Kernel = rbf C = 1.0	0.825
9	10	SVM	Kernel = linear C = 1.0	0.882
9	11	SVM	Kernel = rbf C = 0.1	0.880
9	9	SVM	Kernel = linear C = 1.0	0.871
10	4	RF	Number of trees = 2	0.865
10	6	DT	Tree levels = 2	0.860
10	7	DT	Tree levels = 2	0.845
11	4	SVM	Kernel = rbf C = 0.1	0.917
11	6	SVM	Kernel = rbf C = 1.0	0.913
11	11	SVM	Kernel = rbf C = 1.0	0.910
12	9	RF	Number of trees = 70	0.803
12	10	DT	Tree levels = 2	0.800
12	8	RF	Number of trees = 2	0.785

Analyzing Table 3, it can be observed that the results obtained are really promising, with F1-Score values higher than 0.8 in the best models for every cow. In particular, it is possible to note how the models developed for cows with ID 4 and 6 reached an F1-Score of 1, which indicates a perfect model performance, detecting all estrus records without producing false positives or false negatives. In contrast, models corresponding to cows with ID 8 and 12 showed the worst results, with F1-Score values below 0.85.

The differences in the results obtained between cows can be attributed to three main factors. First, data from some animals may contain more noise, either due to incorrect placement or misalignment of the smart collar, which affects the accuracy of detection of individual behaviors. Second, there may be small errors in the process of sample labeling, which also impacts on the data quality. Finally, some animals may present less defined behavioral patterns, both in their normal activity and in the estrus behavior, which complicates the early detection of this event.

Figure 4 presents a bar graph comparing the best models obtained for each cow, allowing a clear visualization of the performance differences between them.

On the other hand, when analyzing the main classification techniques, it is ob-

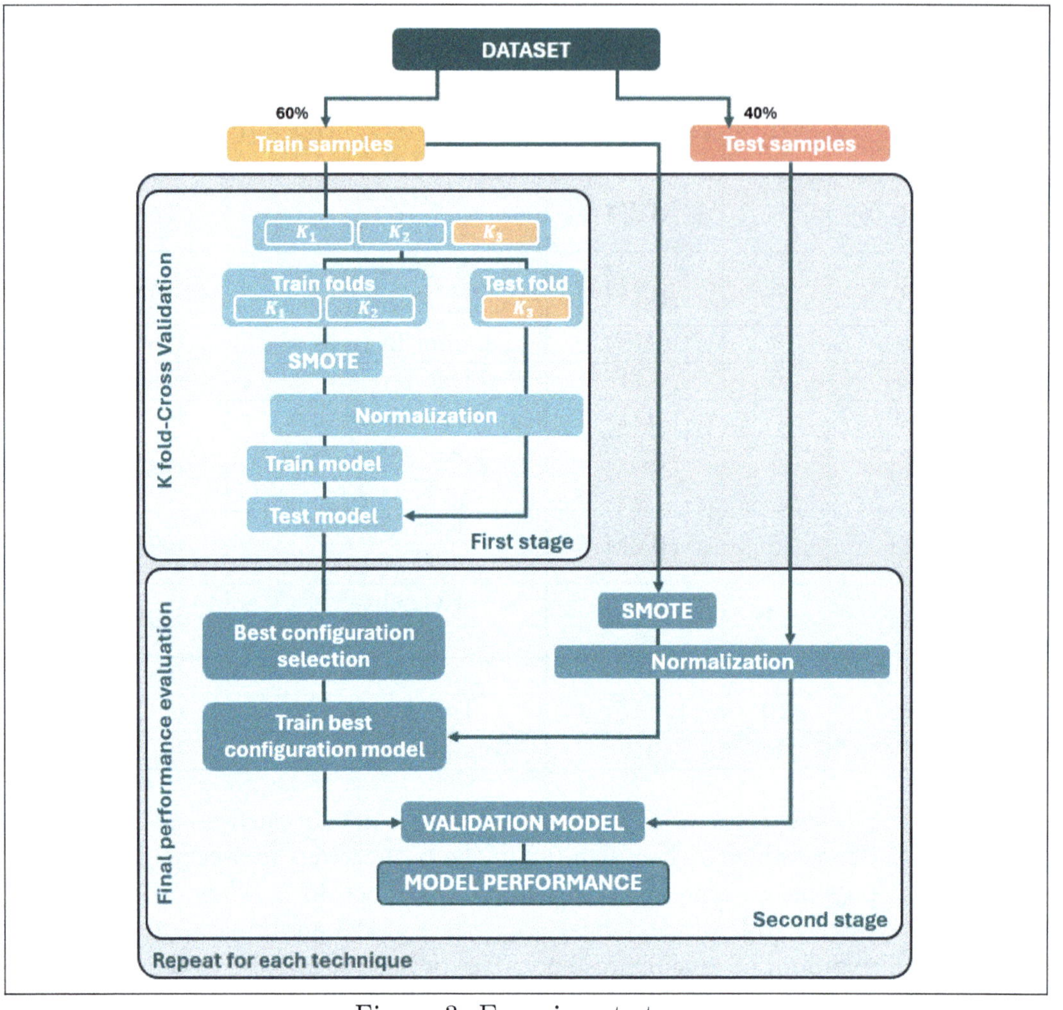

Figure 3: Experiment steps

served that support vector machines show the best performance, together with decision trees, which also achieve outstanding results in several cows. This diversity in the best techniques for each cow highlights substantial differences in the behavioral data for each animal. Therefore, certain techniques may be better adapted to specific behavioral patterns and be more effective on an animal-by-animal basis.

In addition, it is possible to analyze how the performance of the classifiers is influenced by considering a greater number of hours in the generation of the new variables. Figure 5 shows a graph representing the average F1-Score obtained by considering only the best results for each combination of cow and window of hours.

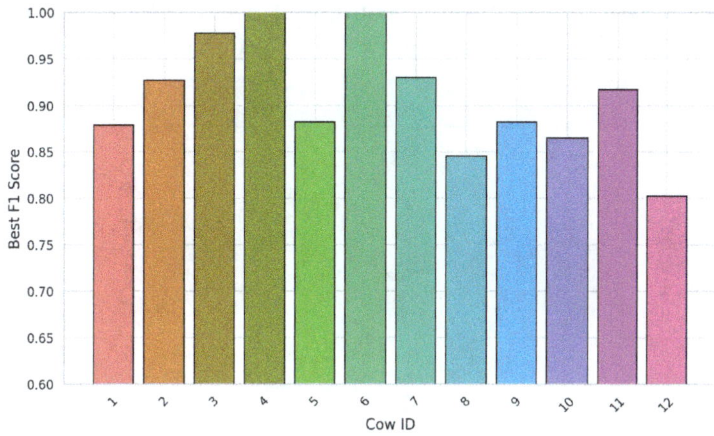

Figure 4: Boxplot of the best results obtained in each cow

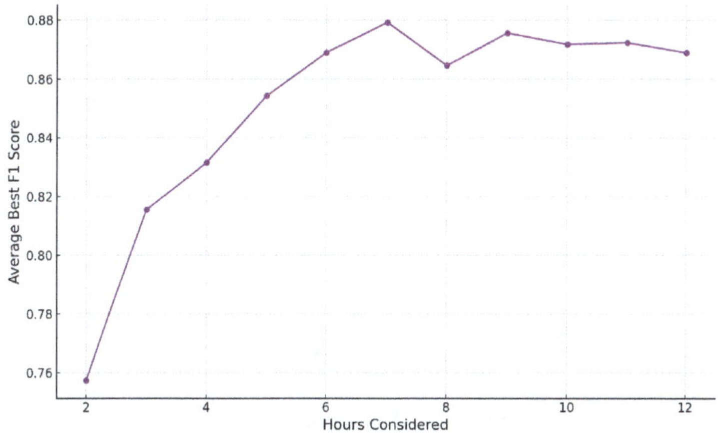

Figure 5: Evolution of the average of the models' performance according to the number of hours considered.

By examining Figure 5, several relevant trends are identified. As the number of hours increases, the average F1-Score improves considerably, peaking around 7 hours. Performance remains high between 7 and 9 hours, with a maximum average F1-Score close to 0.88 when considering 7 hours. However, after 9 hours, although the F1-Score remains high, a slight decrease is observed, suggesting that adding more hours after this point does not contribute significantly to performance and may even worsen it.

In summary, Figure 6 shows a comparative boxplot graph between the results obtained with each technique and time configuration. To obtain this graph, only the best results obtained for each technique were considered, according to the different hour configurations and each animal.

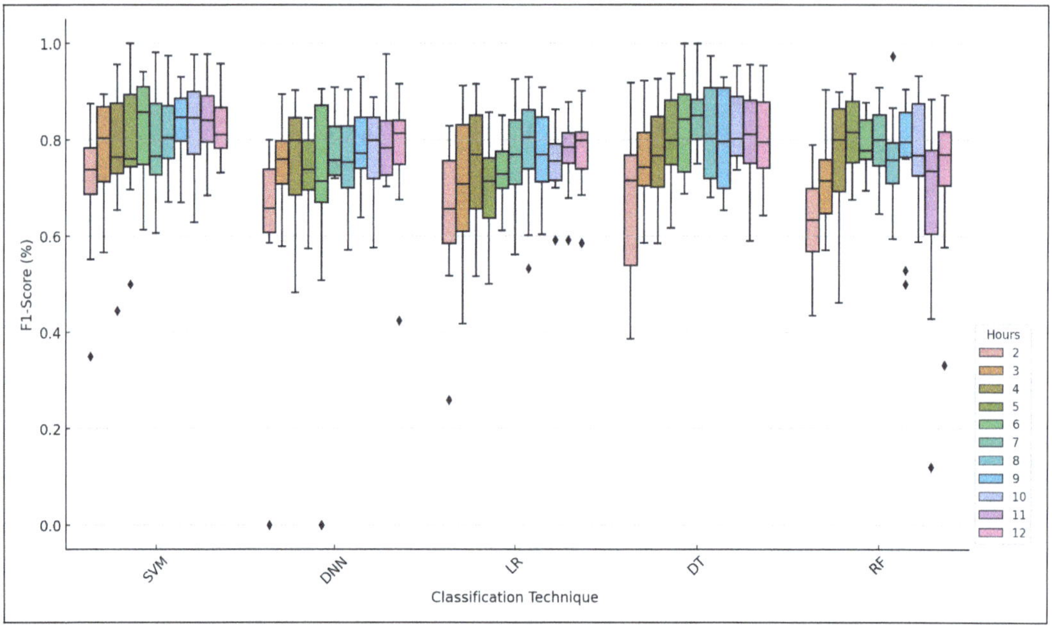

Figure 6: Boxplot of results obtained by technique and hour configuration

6 Conclusions and future research lines

This paper presents an advanced system based on supervised classification techniques for early estrus detection in dairy cattle in intensive farms using behavioral data collected through an intelligent collar. In particular, the RUMI monitoring collar, developed by the Innogando company, has been used. This device allows monitoring of five key activities of the cows and the number of steps.

For this research, data were collected from 12 Holstein Friesian cows of different ages in an intensive dairy farm. From these data, feature engineering techniques were employed to expand the dataset by incorporating new variables that aggregated the duration of each behavior and the number of steps over various time windows, from 2 to 12 previous hours. This variable aggregation approach provided a broader temporal context for the classification models.

Each hour configuration was evaluated using different supervised classification methods, and individualized models were generated for each animal. The results obtained were promising, reaching, in some cases, a perfect classification performance (F1-Score of 1). Support vector machines and decision trees stood out for their efficiency among the techniques employed, obtaining the highest F1-Score scores.

In addition, it was demonstrated that using temporal windows to accumulate behaviors from previous hours significantly improved the models' performance, confirming that temporal context is crucial for accurate estrus detection. In particular, the 7-hour window provided the best performance compared to other settings.

The results also reveal remarkable variability in model performance among different animals, suggesting that behaviors associated with estrus vary considerably among individual cows. This highlights the importance of developing customized models for each animal, given that behavioral patterns can differ significantly. It is also concluded that providing greater temporal context in the data improves accuracy in classifying estrus behaviors.

The tool proposed in this study offers excellent potential for improving dairy cattle management, allowing early detection of estrus and increasing gestation success rates.

As possible lines of future research, further testing is planned with a larger dataset, including a larger number of cows and covering animals from different farms, herds, and breeds. In addition, the performance of this system will be evaluated in extensive regime farms to test its applicability and effectiveness under different management conditions.

Acknowledgments

Álvaro Michelena's research was supported by the Spanish Ministry of Universities (https://www.universidades.gob.es/), under the "Formación de Profesorado Universitario" grant with reference FPU21/00932.

Xunta de Galicia. Grants for consolidating and structuring competitive research units, GPC (ED431B 2023/49).

CITIC, as a center accredited for excellence within the Galician University System and a member of the CIGUS Network, receives subsidies from the Department of Education, Science, Universities, and Vocational Training of the Xunta de Galicia. Additionally, it is co-financed by the EU through the FEDER Galicia 2021-27 operational program (Ref. ED431G 2023/01).

The dataset has been provided by the Galician startup Innogando.

References

[1] E. Trevisi, A. Zecconi, S. Cogrossi, E. Razzuoli, P. Grossi, M. Amadori, Strategies for reduced antibiotic usage in dairy cattle farms, Research in veterinary science 96 (2) (2014) 229–233.

[2] P. Vale, H. Gibbs, R. Vale, M. Christie, E. Florence, J. Munger, D. Sabaini, The expansion of intensive beef farming to the brazilian amazon, Global Environmental Change 57 (2019) 101922.

[3] E. Tullo, A. Finzi, M. Guarino, Environmental impact of livestock farming and precision livestock farming as a mitigation strategy, Science of the total environment 650 (2019) 2751–2760.

[4] D. Darwis, A. R. Mehta, N. E. Wati, S. Samsugi, P. R. Swaminarayan, Digital smart collar: Monitoring cow health using internet of things, in: 2022 International Symposium on Electronics and Smart Devices (ISESD), IEEE, 2022, pp. 1–5.

[5] Á. Michelena, Ó. Fontenla-Romero, J. Luis Calvo-Rolle, A review and future trends of precision livestock over dairy and beef cow cattle with artificial intelligence, Logic Journal of the IGPL (2024) jzae111.

[6] R. Cocco, M. E. A. Canozzi, V. Fischer, Rumination time as an early predictor of metritis and subclinical ketosis in dairy cows at the beginning of lactation: Systematic review-meta-analysis, Preventive Veterinary Medicine 189 (2021) 105309.

[7] S. Büchel, A. Sundrum, Decrease in rumination time as an indicator of the onset of calving, Journal of dairy science 97 (5) (2014) 3120–3127.

[8] M. Stangaferro, R. Wijma, L. Caixeta, M. Al-Abri, J. Giordano, Use of rumination and activity monitoring for the identification of dairy cows with health disorders: Part i. metabolic and digestive disorders, Journal of Dairy Science 99 (9) (2016) 7395–7410.

[9] J. Deming, R. Bergeron, K. Leslie, T. DeVries, Associations of cow-level factors, frequency of feed delivery, and standing and lying behaviour of dairy cows milked in an automatic system, Canadian Journal of Animal Science 93 (4) (2013) 427–433.

[10] G. G. Miguel-Pacheco, J. Kaler, J. Remnant, L. Cheyne, C. Abbott, A. P. French, T. P. Pridmore, J. N. Huxley, Behavioural changes in dairy cows with lameness in an automatic milking system, Applied Animal Behaviour Science 150 (2014) 1–8.

[11] R. C. Chebel, P. R. Silva, M. I. Endres, M. A. Ballou, K. L. Luchterhand, Social stressors and their effects on immunity and health of periparturient dairy cows, Journal of Dairy Science 99 (4) (2016) 3217–3228.

[12] I. Dittrich, M. Gertz, J. Krieter, Alterations in sick dairy cows' daily behavioural patterns, Heliyon 5 (11) (2019) e02902.

[13] S. Reith, S. Hoy, Behavioral signs of estrus and the potential of fully automated systems for detection of estrus in dairy cattle, Animal 12 (2) (2018) 398–407.

[14] A. Carvajal, E. Martínez, M. Tapia, I. T. Ayke, El ciclo estral en la hembra bovina y su importancia productiva, Instituto de investigaciones agropecuarias 246 (2020) 1–4.

[15] J. Roelofs, F. Van Eerdenburg, N. Soede, B. Kemp, Various behavioral signs of estrous and their relationship with time of ovulation in dairy cattle, Theriogenology 63 (5)

(2005) 1366–1377.

[16] U. Riaz, M. Idris, M. Ahmed, F. Ali, L. Yang, Infrared thermography as a potential non-invasive tool for estrus detection in cattle and buffaloes, Animals 13 (8) (2023) 1425.

[17] G. Gautam, Postpartum anestrus in dairy cattle and its management, in: AIP Conference Proceedings, Vol. 2628, AIP Publishing, 2023.

[18] A. Peter, P. Vos, D. Ambrose, Postpartum anestrus in dairy cattle, Theriogenology 71 (9) (2009) 1333–1342.

[19] H. Dobson, R. Smith, M. Royal, C. Knight, I. Sheldon, The high-producing dairy cow and its reproductive performance, Reproduction in domestic animals 42 (2007) 17–23.

[20] J. Roelofs, E. Van Erp-van der Kooij, Estrus detection tools and their applicability in cattle: recent and perspectival situation, Animal Reproduction (AR) 12 (3) (2018) 498–504.

[21] B. Silper, A. Madureira, M. Kaur, T. Burnett, R. Cerri, Comparison of estrus characteristics in holstein heifers by 2 activity monitoring systems, Journal of dairy science 98 (5) (2015) 3158–3165.

[22] V. Schweinzer, E. Gusterer, P. Kanz, S. Krieger, D. Süss, L. Lidauer, A. Berger, F. Kickinger, M. Öhlschuster, W. Auer, et al., Evaluation of an ear-attached accelerometer for detecting estrus events in indoor housed dairy cows, Theriogenology 130 (2019) 19–25.

[23] S. Benaissa, F. A. M. Tuyttens, D. Plets, J. Trogh, L. Martens, L. Vandaele, W. Joseph, B. Sonck, Calving and estrus detection in dairy cattle using a combination of indoor localization and accelerometer sensors, Computers and electronics in agriculture 168 (2020) 105153.

[24] Z. Wang, S. Wang, C. Wang, Y. Zhang, Z. Zong, H. Wang, L. Su, Y. Du, A non-contact cow estrus monitoring method based on the thermal infrared images of cows, Agriculture 13 (2) (2023) 385.

[25] P. Bruyère, T. Hétreau, C. Ponsart, J. Gatien, S. Buff, C. Disenhaus, O. Giroud, P. Guérin, Can video cameras replace visual estrus detection in dairy cows?, Theriogenology 77 (3) (2012) 525–530.

[26] Innogando web page, `https://innogando.com/`, accessed: 2024-08-05 (2024).

[27] S. Sperandei, Understanding logistic regression analysis, Biochemia medica 24 (1) (2014) 12–18.

[28] D. W. Hosmer Jr, S. Lemeshow, R. X. Sturdivant, Applied logistic regression, Vol. 398, John Wiley & Sons, 2013.

[29] B. Charbuty, A. Abdulazeez, Classification based on decision tree algorithm for machine learning, Journal of Applied Science and Technology Trends 2 (01) (2021) 20–28.

[30] B. Gupta, A. Rawat, A. Jain, A. Arora, N. Dhami, Analysis of various decision tree algorithms for classification in data mining, International Journal of Computer Applications 163 (8) (2017) 15–19.

[31] A. Parmar, R. Katariya, V. Patel, A review on random forest: An ensemble classifier, in:

International conference on intelligent data communication technologies and internet of things (ICICI) 2018, Springer, 2019, pp. 758–763.

[32] R. Saravanan, P. Sujatha, A state of art techniques on machine learning algorithms: a perspective of supervised learning approaches in data classification, in: 2018 Second international conference on intelligent computing and control systems (ICICCS), IEEE, 2018, pp. 945–949.

[33] C. Cortes, V. Vapnik, Support-vector networks, Machine learning 20 (3) (1995) 273–297.

[34] W. S. Noble, What is a support vector machine?, Nature biotechnology 24 (12) (2006) 1565–1567.

[35] J. Cervantes, F. Garcia-Lamont, L. Rodríguez-Mazahua, A. Lopez, A comprehensive survey on support vector machine classification: Applications, challenges and trends, Neurocomputing 408 (2020) 189–215.

[36] I. N. Da Silva, D. H. Spatti, R. A. Flauzino, L. H. B. Liboni, S. F. dos Reis Alves, Artificial neural networks, Cham: Springer International Publishing 39 (2017). doi: 10.1007/978-3-319-68612-7.

[37] O. I. Abiodun, A. Jantan, A. E. Omolara, K. V. Dada, N. A. Mohamed, H. Arshad, State-of-the-art in artificial neural network applications: A survey, Heliyon 4 (11) (2018).

[38] M. Cilimkovic, Neural networks and back propagation algorithm, Institute of Technology Blanchardstown, Blanchardstown Road North Dublin 15 (1) (2015).

[39] R. Mohammed, J. Rawashdeh, M. Abdullah, Machine learning with oversampling and undersampling techniques: overview study and experimental results, in: 2020 11th international conference on information and communication systems (ICICS), IEEE, 2020, pp. 243–248.

[40] N. V. Chawla, K. W. Bowyer, L. O. Hall, W. P. Kegelmeyer, Smote: synthetic minority over-sampling technique, Journal of artificial intelligence research 16 (2002) 321–357.

[41] S. Patro, Normalization: A preprocessing stage, arXiv preprint arXiv:1503.06462 (2015).

[42] Á. Michelena, J. Aveleira-Mata, E. Jove, M. Bayón-Gutiérrez, P. Novais, O. F. Romero, J. L. Calvo-Rolle, H. Aláiz-Moretón, A novel intelligent approach for man-in-the-middle attacks detection over internet of things environments based on message queuing telemetry transport, Expert Systems 41 (2) (2024) e13263.

 Received 4 November 2024

Detection and Classification of DoS Attacks in Zigbee Networks Using Supervised Learning Algorithms

Jose Aveleira-Mata and María Teresa García-Ordás
SECOMUCI Research Group, Department of Electric, Systems and Automatics Engineering, Universidad de Leon. Campus de Vegazana, s/n, 24071, León, Spain
{jose.aveleira, mgaro}@unileon.es

Marta-María Álvarez-Crespo and Esteban Jove
University of A Coruña, CTC, CITIC, Department of Industrial Engineering. Calle Mendizábal s/n, 15403, Ferrol, A Coruña, Spain.
{esteban.jove,marta.maria.alvarez.crespo}@udc.es

Míriam Timiraos
University of A Coruña, CTC, CITIC, Department of Industrial Engineering. Calle Mendizábal s/n, 15403, Ferrol, A Coruña, Spain and Fundación Instituto Tecnológico de Galicia, Department of Water Technologies, National Technological Center, 15003, A Coruña, Spain
miriam.timiraos.diaz@udc.es

Héctor Alaiz-Moretón
4RIASC. Research Institute of Applied Sciences in Cybersecurity. Universidad de Leon. MIC. Campus de Vegazana, s/n, 24071, León, Spain.
{hector.moreton}@unileon.es

Abstract

The Zigbee protocol, designed for low-power wireless personal area networks, is a key technology for the Internet of Things (IoT). This paper explores detecting denial-of-service (DoS) attacks in Zigbee networks through supervised classification methods. A specialized dataset was generated from a simulated Zigbee environment, capturing both normal and malicious traffic. The study evaluates the effectiveness of six supervised classification algorithms, including Logistic

Regression, K-Nearest Neighbors (KNN), Support Vector Machines (SVM), Decision Trees, Random Forests, and Artificial Neural Networks (ANN). Results indicate that KNN, Decision Trees, and Random Forests achieved superior performance with high accuracy and efficiency. This methodology highlights the potential of tailored datasets and machine learning approaches for enhancing security in Zigbee networks and protecting IoT environments from evolving threats.

1 Introduction

The Internet of Things (IoT) enables connectivity among objects, enhancing their functionalities and interactions with the surrounding environment. Through sensors and actuators, IoT systems gather environmental data and trigger actions, respectively, creating adaptive responses and facilitating automation across diverse sectors. This capability has led to significant advancements in industries such as smart cities, manufacturing, residential automation, and healthcare, where IoT devices play an essential role in optimizing operations, monitoring environments, and improving user experiences [14, 28]. The increasing adoption of IoT devices stems from their ability to streamline operations and support data-driven decision-making processes, making them a cornerstone for digital transformation initiatives

According to estimates, the number of IoT connected devices worldwide is expected to reach 75.44 billion by 2025 [16].

IoT systems rely on various network protocols and technologies to meet specific communication needs. Some key protocols include Zigbee, Z-Wave, LoRa, and Bluetooth Low Energy (BLE). These protocols follow the IEEE 802.15.4 standard, which defines the basics of low-rate wireless personal area networks (LR-WPANs) and supports reliable, low-power communications for devices with limited energy resources [13]. Zigbee and Z-Wave are commonly used for home automation due to their low power use and ability to create mesh networks, enhancing range and coverage. LoRa, with its long-range and low data rate, is ideal for applications like environmental monitoring, while BLE enables short-range, high-speed connections suitable for wearable devices and healthcare applications [25, 32]

However, the growth of IoT systems also brings complex cybersecurity challenges. The diversity of devices and the wide range of communication protocols increase possible points of attack, creating vulnerabilities that malicious actors can exploit. Common attacks include data interception (sniffing), message manipulation, denial-of-service (DoS) attacks, and physical security breaches [11]. Additionally, man-in-the-middle (MITM) attacks, where attackers intercept and alter communication between devices, pose serious threats to data integrity, especially in applications like healthcare and smart cities [24]. Furthermore, botnet attacks—where a network of compromised IoT devices is used to overwhelm servers—are becoming increasingly common, making IoT devices a target for cybercriminals seeking to disrupt services or steal information [21].Implementing strong security measures is difficult because many IoT devices have limited processing power, storage, and battery life. These constraints make it hard to use traditional security solutions and call for innovative approaches to protect data privacy and security in IoT systems.

Recent research underscores the urgency of developing autonomous and resource efficient systems for attack detection in IoT devices with limited capacity, emphasizing the role of Intrusion Detection Systems (IDS) as a practical solution [3]. These systems, especially machine learning-based Anomaly Detection Systems (ADS), are seen as effective because they operate without needing modifications to existing IoT infrastructures. ADS use machine learning models trained on specialized datasets to identify unusual patterns in device behavior, enabling proactive attack detection within current IoT setups. A systematic review by [15] highlights the importance of these specialized datasets, showing how they can enhance the accuracy of IDS models in recognizing threats specific to IoT environments. Research by [36, 12] further examines the performance of intelligent IDS systems in classifying and detecting both normal and malicious traffic within IoT networks that rely on Wi-Fi. This approach preserves the integrity of existing IoT systems while providing a pathway to more resilient, adaptive security measures against evolving cyber threats.

This study focuses on enhancing the security of the Zigbee protocol by addressing specific vulnerabilities, particularly those related to AT command attacks [29]. To achieve this, we created an IoT environment employing Zigbee, where we simulated a Denial-of-Service (DoS) attack using AT commands. All traffic generated within this environment was captured and labeled to construct a dataset containing both normal and anomalous traffic. This dataset was then used to train several supervised classification models, ultimately producing a binary model for network traffic analysis and DoS attack detection within Zigbee networks. In addition, recent work has shown that intrusion detection systems (IDS) trained on Zigbee-specific datasets can significantly improve detection accuracy in identifying malicious commands and DoS attacks, reinforcing the importance of tailored datasets for protocol-specific security measures [2].

To address the imbalance in the dataset, the SMOTE (Synthetic Minority Oversampling Technique) method is employed to generate synthetic samples and ensure a more balanced class distribution. Additionally, six established soft computing techniques are utilized: K-Nearest Neighbors (KNN), Logistic Regression (LR), Random Forest (RF), Decision Trees (DT), Artificial Neural Networks (ANN), and Support Vector Machines (SVM). These methods are applied to perform analysis and classification tasks, providing a robust framework for evaluating performance and suitability across various scenarios.

This work is organized as follows: Section 2 introduces the case study, providing details on the deployment of the test environment and an in-depth description of the dataset utilized in this research. Section 3 presents the techniques used. Section 4 outlines the parameter configurations for each of the methods discussed. In Section 5, several performance metrics are introduced alongside a discussion identifying the most effective method among those evaluated. Finally, Section 6 concludes the manuscript and proposes directions for future research.

2 Case study

Zigbee is a low-power wireless communication protocol widely used in the Internet of Things (IoT) to support applications in Wireless Personal Area Networks (WPANs). Developed to enable effective data exchange in devices with limited power resources, Zigbee's network architecture includes mesh, star, and tree topologies, which provide flexibility in configuring different network layouts depending on application needs [26]. This adaptability allows for the integration of various devices across smart home systems, industrial automation, and healthcare monitoring applications. The protocol is based on the IEEE 802.15.4 standard, which underpins its efficient low-rate, short-range communication capabilities. Due to its low energy consumption and compatibility with a range of IoT devices, Zigbee is often preferred for creating interconnected environments where devices can reliably communicate with minimal maintenance[4]

To create an environment that simulates real Zigbee traffic, we established a configuration using XBee modules [9] designed to operate under the Zigbee protocol. In this setup, one XBee module functions as an end device and is connected to a DHT22 sensor, which measures temperature and humidity. This connection is achieved through the input/output pins provided on the XBee module, allowing real-time data collection from the sensor. A second XBee module is configured as a router; it acts as a relay point, receiving and passing data within the Zigbee network. This router is paired with an ESP8266 chip, which serves as a gateway device. The ESP8266 connects the Zigbee network to the internet via Wi-Fi, sending the sensor data to a remote server for analysis and storage. Additionally, a third XBee module is configured to act as the network coordinator, managing communication paths and ensuring reliable data transmission between devices. This configuration emulates a standard Zigbee network structure, with three core components: an end device for data sensing, a central coordinator to manage and organize data flow, and a router

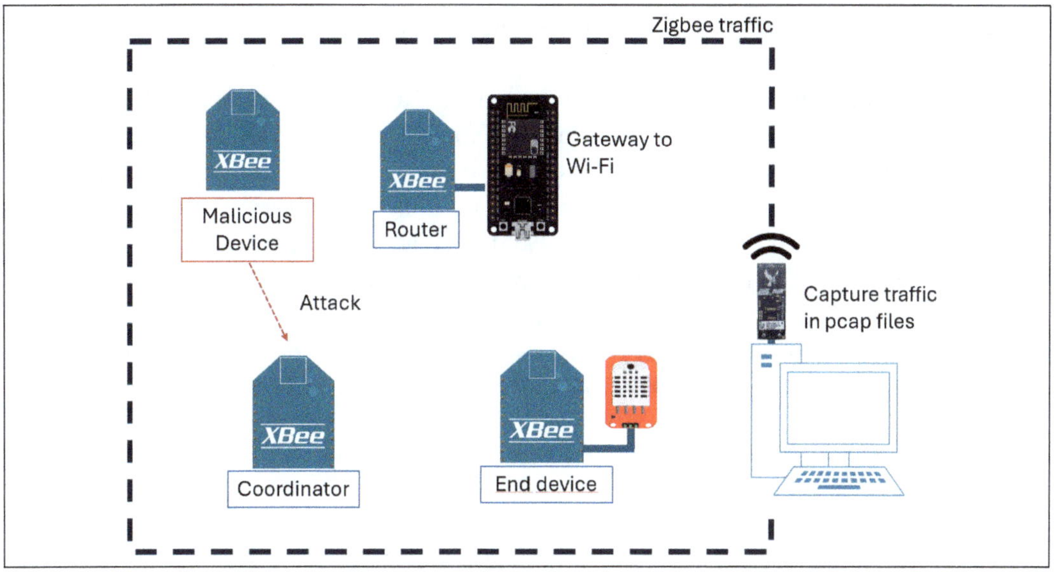

Figure 1: Zigbee enviroment

to extend network range and support data forwarding.

All network traffic generated by the Zigbee system is captured to create a dataset containing both normal and attack-labeled frames. In this setup, the Rzusbstik antenna is used as the capture device. To enable this, the Rzusbstik's firmware needs to be modified by installing the KillerBee firmware [27], which is specialized for Zigbee protocol analysis. This firmware modification allows the antenna to operate in promiscuous mode, a state where it can capture all packets within range, regardless of their destination MAC address.Promiscuous mode allows packet capture tools to monitor all network traffic, facilitating the identification of potential intrusions and irregular traffic patterns

In the context of this environment, a denial of service (DoS) attack is executed that compromises the functionality of the router, while the coordinator attempts to restore the network. Although the router plays a crucial role as a gateway for data transmission to the Internet, allowing it to be displayed in client applications, the Zigbee nodes continue to operate under the management of the coordinator. ,the diagram of this attaco to the environment can be seen in 1.

The attack process involves the infiltration of a malicious device into the target Zigbee network, ensuring it operates on the same channel and masquerades as a coordinator. This process begins with a reconnaissance phase, during which the attacker uses sniffing techniques to scan the Zigbee network by broadcasting probe requests across all channels. Upon detecting a response, typically in the form of a beacon packet from nearby devices, the attacker gains access to essential network information, such as the PANID (Personal Area Network Identifier). Armed with the PANID, the attacker can insert a malicious node into the network. Within this compromised network, the attacker can execute specialized AT commands to remotely manipulate the device configurations [29]. By altering key parameters, such as the network ID and the Router Enable (RE) setting, the attacker is able to isolate and exclude the legitimate router from the network. This effectively disrupts the router's services, preventing it from transmitting critical data, like temperature readings, to the Internet server. This approach not only disables the router but also compromises the integrity and availability of the data flow within the Zigbee network.

In the described environment, all network activity (both during normal operation and while under attack) is captured using the Rzusbstick antenna, as detailed previously. Network traffic is recorded and stored in PCAP (Packet Capture) files, a widely accepted format for network packet analysis. To facilitate further analysis, these PCAP files are converted to CSV format, which allows for easier manipulation and integration with data analysis tools. This conversion process includes extracting all fields relevant to the Zigbee protocol, as specified in the "Wireshark Display Filter Reference: Zigbee Network Layer" section [33]. Each captured frame is subsequently labeled to indicate whether it was recorded during a DoS attack or represents normal traffic, taking into account the specific timestamps that mark the start and end of each attack phase. This labeling enables the identification and separation of attack patterns from normal network behavior, providing a foundation for subsequent security analyses and model training

The dataset consists of N = 39,387 Zigbee frames captured in CSV format using Wireshark/TShark. It includes 144 attributes per sample, of which 119 are numerical (e.g., frame.time_delta [0–17.627 s], frame.time_relative [0–3,552.753 s], frame.len [50–225 bytes], Zigbee-level fields like zbee_nwk.dst_address [0–65535], etc.) and 25 are categorical (e.g., ip.src, frame.interface_id, Zigbee flags, all one-hot encoded). The dataset also includes a type label with two possible values: normal and attack. The original dataset is composed of 26,201 normal samples (66.52%) and 13,186 DoS attack samples (33.48%). To correct this imbalance, SMOTE was applied to the minority class, resulting in a balanced dataset of 52,402 samples, equally split between normal and attack traffic. This version was used for model training and evaluation to ensure fair performance across both classes.

3 Impemented methods

This section outlines the implemented methods. The following provides a detailed explanation of the techniques and approaches applied in this work.

3.1 Data Balancing

In supervised classification, an imbalanced dataset, where one class is underrepresented, can lead to models favoring the majority class, which is problematic in critical applications. Techniques like Synthetic Minority Oversampling Technique help balance the class distribution, improving model performance.

3.1.1 Synthetic Minority Oversampling Technique (SMOTE)

Class imbalance is a significant challenge in supervised classification tasks, particularly when the minority class is underrepresented in the dataset. This imbalance often results in biased models that favor the majority class, overlooking the critical patterns of the minority class. SMOTE has emerged as a widely recognized method for balancing datasets to address this issue. Introduced by [8], SMOTE generates synthetic samples for the minority class, improving its representation without simply duplicating existing data points. This approach enhances the learning process and supports better model performance on imbalanced datasets.

SMOTE operates by creating new data points through interpolation between existing samples of the minority class. Unlike traditional oversampling techniques that merely replicate data, SMOTE ensures the creation of diverse and realistic synthetic samples. This is achieved by selecting a sample from the minority class, identifying its nearest neighbors, and generating a new sample along the line connecting the original point and one of its neighbors. By systematically repeating this process, SMOTE increases the presence of the minority class, effectively reducing the skewed distribution of the dataset [8, 30].

A critical advantage of SMOTE lies in its ability to address overfitting, a common issue with standard oversampling methods. Replicating data points can lead to models that overly rely on specific patterns in the training set, limiting their ability to generalize. In contrast, the synthetic data created by SMOTE introduces variability, fostering a more robust learning process. Furthermore, the technique is flexible and can be paired with various supervised classification algorithms, including Logistic Regression, Support Vector Machines (SVM), and k-Nearest Neighbors (KNN), to enhance their predictive accuracy when dealing with imbalanced data [8].

Despite its strengths, SMOTE is not without limitations. One challenge arises when synthetic samples are generated in regions where the classes overlap. This can blur the boundaries between classes, potentially leading to misclassification. Additionally, SMOTE's effectiveness is limited in scenarios where the minority class is extremely small, as the lack of sufficient original samples may hinder the creation of meaningful synthetic points. Moreover, the oversampling process increases the size of the dataset, which can escalate computational costs during model training, particularly for algorithms that are sensitive to data volume [8].

Over time, several extensions of SMOTE have been developed to mitigate these drawbacks. For instance, Borderline-SMOTE focuses on generating synthetic samples near the decision boundaries, emphasizing areas that are more likely to lead to misclassifications. Similarly, SMOTE-NC adapts the technique for datasets containing both numerical and categorical variables. Another variation, ADASYN (Adaptive Synthetic Sampling), prioritizes generating synthetic samples for minority class instances that are harder to classify, further refining the approach [8, 31].

By addressing the imbalance between classes, SMOTE ensures that supervised classification models can learn from minority class instances without bias. This not only improves model performance but also enhances its reliability in real-world scenarios where minority classes often represent critical events or outcomes [8, 30, 31].

3.2 Supervised classification algorithms

Currently, various advanced methods are used to address classification problems. However, this study focuses on three simple yet effective supervised classification techniques: Logistic Regression, K-Nearest Neighbors (KNN), and Support Vector Machines (SVM). These algorithms were selected due to their demonstrated effectiveness in previous cybersecurity studies [10, 19, 18]. Below, a brief overview of each technique is provided.

3.2.1 Logistic Regression

Logistic Regression (LR) is a widely used supervised learning technique for addressing classification problems [7]. Unlike linear regression, which focuses on predicting continuous outcomes, logistic regression is designed to estimate the probability of a categorical outcome based on given predictor variables. It achieves this by utilizing the logistic or sigmoid function to map input values into a range between 0 and 1. The logistic function, or sigmoid function, is defined as Ecuation1, where (z) is the linear combination of input features. [5]

$$\sigma(z) = \frac{1}{1 + e^{-z}} \tag{1}$$

During training, logistic regression optimizes the model parameters (weights and bias) to minimize the difference between the predicted probabilities and the actual class labels. This is typically achieved using the binary cross-entropy loss function, Ecuation 2, where (y) is the true label, ($\hat{y}$) is the predicted probability, and (N) is the number of samples [5].

$$L(y, \hat{y}) = -\frac{1}{N} \sum_{i=1}^{N} [y_i \log(\hat{y}_i) + (1 - y_i) \log(1 - \hat{y}_i)] \tag{2}$$

This method calculates the likelihood of a specific event or class occurring by applying the logistic function to the data. Based on this probability, it classifies the input into the most probable category, provided the probability exceeds a defined threshold, typically 0.5. This threshold can be adjusted depending on the specific requirements of the problem. To prevent overfitting, regularization techniques such as L1 (Lasso) and L2 (Ridge) regularization can be applied. These techniques add a penalty term to the loss function, which discourages the model from fitting the noise in the training data [7].

Logistic regression is highly popular due to its simplicity and interpretability. Additionally, it provides insights into the impact of each predictor variable on the probability of the outcome, making it a valuable tool for analyzing and predicting categorical data. Logistic regression has several advantages, including being simple to implement and interpret, computationally efficient, and providing probabilities for class membership. However, it also has limitations, such as assuming a linear relationship between the input features and the log-odds of the outcome, and it may not be suitable for complex relationships unless combined with other techniques [5, 7].

3.2.2 K-Nearest Neighbors

The k-Nearest Neighbor (KNN) algorithm is a non-linear supervised classification technique that classifies data points based on the principle that points within the same class tend to be close to each other in the feature space. The classification of a new data point is determined by the dominant class among its k nearest neighbors [7].

The performance of the KNN algorithm relies heavily on selecting an appropriate value for k. Fine-tuning this parameter is essential, often involving strategies such as testing performance across different k values to identify the optimal choice for the specific dataset. However, larger k values can result in overly generalized decision boundaries.

Unlike many other classification methods, KNN does not build a model during the training phase. Instead, it stores the entire training dataset, which enables a fast training process. However, classifying a new data point can be computationally intensive, as it requires calculating the distance from the new point to all stored points. This can pose challenges for applications where quick classification is critical. To determine the nearest neighbors, various distance metrics are employed, with Euclidean distance being the most commonly used. The Euclidean distance between two points (x) and (y) in an (n)-dimensional space is given by Ecuation 3 [35].

$$d(x, y) = \sqrt{\sum_{i=1}^{n}(x_i - y_i)^2} \tag{3}$$

Other distance metrics, such as Manhattan distance and Minkowski distance, can also be used depending on the nature of the data and the specific requirements of the application. The choice of distance metric can significantly impact the performance of the KNN algorithm. Additionally, the algorithm's sensitivity to the local structure of the data means that feature scaling or normalization is often necessary to ensure that all features contribute equally to the distance calculations. This is particularly important when the features have different units or scales.
KNN is a versatile algorithm used in various applications, including pattern recognition, data mining, and intrusion detection. Its simplicity and effectiveness make it a popular choice for both classification and regression tasks. However, its computational complexity and sensitivity to irrelevant features or noise in the data are notable limitations [7, 35].

3.2.3 Support Vector Machines

Support Vector Machines (SVMs) are supervised learning models introduced by Cortes and Vapnik, widely applied in both classification and regression tasks. The core goal of SVMs is to determine a hyperplane that maximizes the margin, or the minimum distance, between the hyperplane and the closest data points from each class. These points, known as support vectors, are critical for defining the decision boundary [17].

Traditional SVMs assume that data classes can be separated using a linear boundary. However, real-world datasets are often more complex and may not be linearly separable. To address this limitation, SVMs use a transformation function to map the data into a higher-dimensional space where linear separation is feasible. This transformation is performed using a *kernel* function, denoted as $\phi(x)$, which determines how the data is projected into the new space.

The optimization problem for finding the optimal hyperplane can be formulated as follows: given a set of training examples $((x_i, y_i))$ where (x_i) represents the feature vectors and (y_i) represents the class labels, the goal is to find the weight vector (w) and bias (b) that minimize the objective function 4 [34].

$$\frac{1}{2}|w|^2 \tag{4}$$

Subject to the constrains in Ecuation 5.

$$y_i(w \cdot x_i + b) \geq 1 \quad \text{for all } i \tag{5}$$

This is a quadratic optimization problem that can be solved using methods such as Lagrange multipliers. The solution involves only the support vectors, which are the data points closest to the decision boundary. These support vectors are crucial because they define the position and orientation of the hyperplane [34].

In cases where the data is not linearly separable, SVMs employ the kernel trick to transform the data into a higher-dimensional space. Commonly used kernel functions include the polynomial kernel, the radial basis function (RBF) kernel, and the sigmoid kernel. The kernel function choice depends on the dataset's specific characteristics and the problem at hand [34].

SVMs are known for their robustness and effectiveness in high-dimensional spaces, making them suitable for various applications, including text classification, image recognition, and bioinformatics. Despite their advantages, SVMs can be computationally intensive, especially with large datasets, and require careful tuning of parameters such as the regularization parameter (C) and the kernel parameters [34].

3.2.4 Decision Trees

Decision Trees (DT) are a versatile, non-parametric method in supervised machine learning, often employed for both classification and regression tasks [6]. This technique builds models in a tree-like structure comprising a root node, branches, decision nodes, and terminal nodes known as leaves. The process begins at the root node, where a variable from the dataset is evaluated. Based on its value, the data is routed along one of the branches to the next decision node. This progression continues until the data reaches a leaf node, which assigns the final class or prediction [22].

Decision nodes function as splitting points, where specific variables are analyzed to determine how the data should be divided, while leaf nodes provide the classification or outcome. Since the tree's performance heavily depends on the variables selected at each stage, careful selection of splitting criteria is crucial [22, 6].

To determine the optimal splits, the method quantifies data impurity at each node using measures such as entropy or the Gini index. These metrics evaluate the level of disorder or impurity in the dataset. Information gain, a value derived by comparing the impurity before and after a split, helps prioritize variables. Variables with the highest information gain are placed in the tree's upper levels, as these splits reduce impurity the most. The tree is then constructed sequentially from the root downward, ensuring the most significant splits occur early in the process [22].

A key strength of decision trees lies in their interpretability. Their hierarchical structure is easy to understand, and the splitting process can be visualized, making them highly accessible to users. However, this simplicity comes with a drawback: decision trees can overfit data, especially in complex scenarios. Overfitting occurs when the model becomes overly tailored to the training data, resulting in large and unwieldy trees that fail to generalize effectively to new datasets [6].

Despite this limitation, decision trees remain popular due to their balance of simplicity and utility, particularly in applications where interpretability and clear decision-making processes are essential [22, 6].

3.2.5 Random Forest

Random Forest (RF) is a robust supervised machine learning algorithm commonly utilized for classification tasks and adaptable for regression problems as well. The method leverages an ensemble of decision trees to create a more accurate and stable predictive model [23]. In this approach, multiple decision trees are independently trained, and the final classification result is determined by majority voting, where the class predicted most frequently by the individual trees becomes the model's output [1].

Each decision tree within the random forest is trained on a unique subset of the training dataset. These subsets are generated using a technique called Bootstrap Aggregation, or Bagging. Bagging involves randomly sampling data points with replacements to create diverse training subsets, allowing the same data point to appear in multiple subsets. This ensures variability among the trees, which enhances the model's generalization capabilities [23].

The construction of a random forest involves two key steps. First, the Bagging method produces as many random subsets as there are decision trees in the forest. Each tree is trained independently on its respective subset, using a random selection of features at each split to reduce the correlation between the trees. Once all the trees are trained, the random forest aggregates their predictions to classify the input data. For classification tasks, the class predicted most frequently by the ensemble is selected as the output, while for regression tasks, the average prediction of all trees is computed [1].

RF's ability to handle large datasets and its resistance to overfitting make it a widely used algorithm in machine learning. It is especially advantageous in applications where model accuracy and robustness are critical [1, 23].

3.2.6 Artificial Neural Networks

Artificial Neural Networks (ANNs) are computational models inspired by the structure and functionality of biological neural networks found in the human brain. ANNs consist of artificial neurons that are interconnected to form a network or matrix. These neurons simulate basic brain functions, processing input data through a series of mathematical operations to produce an output [7].

An artificial neuron operates in three main steps. First, each input is assigned a weight—a numerical value that determines the importance of the input. The weighted inputs are then summed together along with a bias term, an independent value added to the calculation to adjust the result. Finally, this sum is passed through an activation function, which transforms it into the neuron's output. Activation functions vary depending on the application and may include options like linear, sigmoid, or ReLU (Rectified Linear Unit) [20].

Neurons are organized into layers, which define the network's structure. The most common configuration is the Multilayer Perceptron (MLP), a type of ANN composed of an input layer, one or more hidden layers, and an output layer. The input layer receives the data, the hidden layers process it, and the output layer generates the final prediction or classification [7].

Training an ANN involves adjusting the weights and biases in each layer to minimize prediction errors. This process is typically carried out using backpropagation, a method that calculates gradients to optimize the network's weights efficiently. To prevent overfitting, techniques such as dropout layers are employed. A dropout layer randomly "ignores" certain neurons during training, effectively reducing the network's complexity and improving its generalization ability. The influence of dropout can be controlled using a probability parameter [20].

ANN are highly adaptable and can be configured in numerous ways by varying the number of layers, the number of neurons per layer, and the connections between them. This flexibility makes ANNs a powerful tool for solving a wide range of problems, from image recognition to predictive modeling, while their structure and training methods allow for continual refinement to achieve optimal performance [7, 20].

4 Experiment configuration

To perform the experiments, a Python algorithm was developed to integrate the three supervised classification techniques previously described. The experimental methodology is detailed below, focusing on dataset preprocessing, hyperparameter tuning, model validation, and result comparison. The steps followed for the experiment setup are illustrated in Figure 2 and are explained in detail below.

Data Preprocessing: The dataset underwent extensive preprocessing to optimize classifier performance. This process included converting categorical variables to numerical values, eliminating constant predictor variables, removing those with more than 20% missing data, and excluding variables that showed a correlation greater than 0.9 with other variables. Additionally, to prevent overfitting, duplicate samples were removed.

Data Splitting and Normalization: The dataset was then split into 70% for training and 30% for validation. The training data were normalized using the z-score method, which adjusts the data to have a mean of 0 and a standard deviation of 1.

Hyperparameter tuning: Hyperparameters for each classification technique were tuned by evaluating different configurations through cross-validation using 10 data subsets. Given the imbalance in the dataset, the SMOTE method was employed to balance the classes. SMOTE works by generating synthetic samples for the minority class by interpolating between existing minority class samples. This helps to create a more balanced dataset, ideally achieving a 50/50 distribution between the classes. By doing so, the performance of the classifiers can be improved, as they are provided with more representative training data.

The configurations tested for each method are listed in Table 1.

Model evaluation: For the different techniques used, the models obtained with each configuration were evaluated. To compare the models, the Kruskal-Wallis analysis of variance and Tukey's method were applied, allowing for a statistical evaluation of the differences between the models. The Area Under the Curve (AUC) was used as a comparison metric, as it is insensitive to the imbalance between classes.

Training and Final Validation: With the best configurations defined for each method, each model was trained using the entire training dataset and validated with the validation data (30% of the initial dataset).

5 Results

In this study, each model's performance was evaluated using the Area Under the Curve (AUC) as the primary comparison metric and the F1-score as an additional measure of accuracy. The F1-score, like the AUC, ranges from 0 to 1, where a

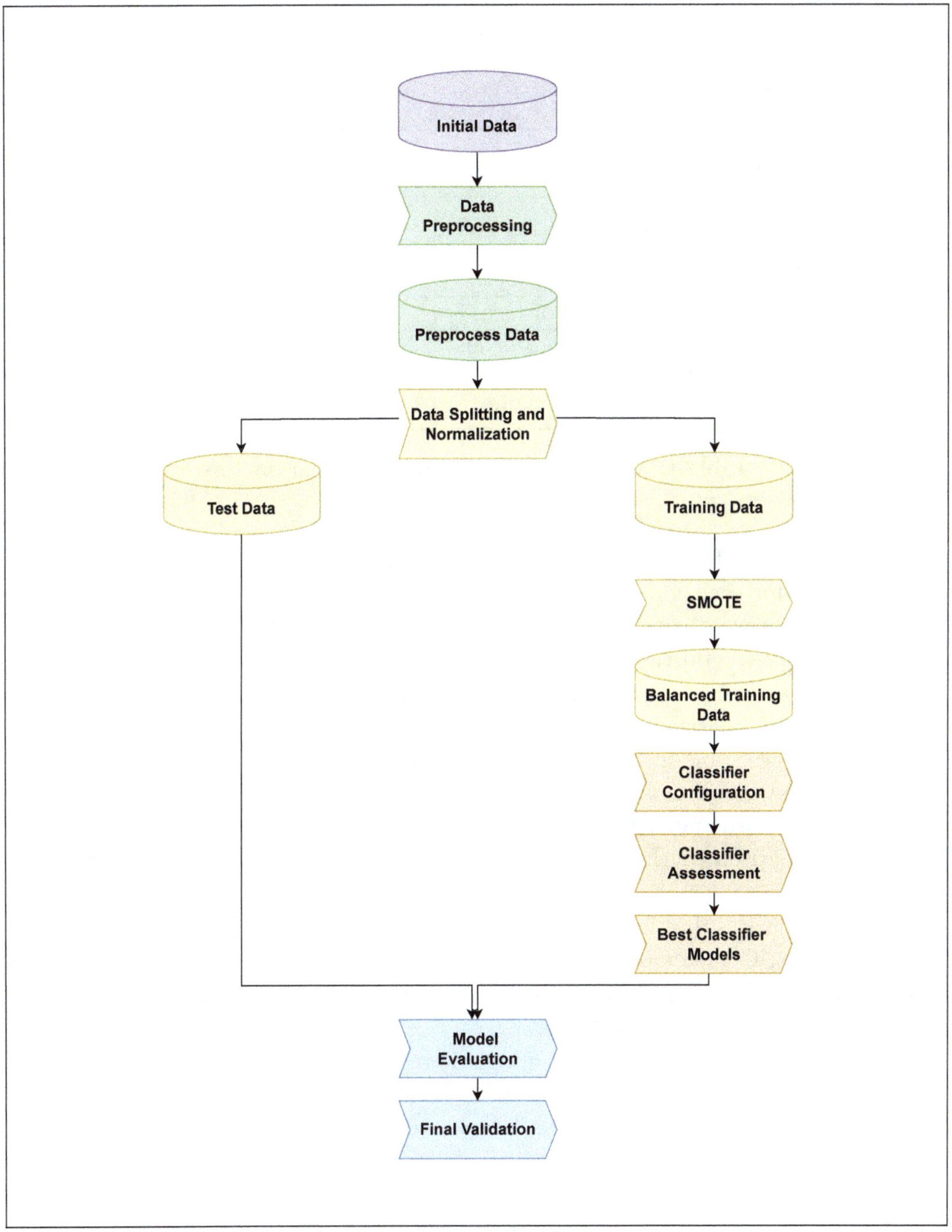

Figure 2: Experiment's setup.

Evaluated technique	Evaluated configuration	Hyperparameters values
KNN	Number of neighbor	5, 10, 25, 50, 100, 250, 500 and 1000
SVM	Kernel Regularization (C)	linear, poly and rbf 1, 0.1 and 0.01
LR	Regularization (C)	0.0001, 0.0005, 0.001, 0.005, 0.01, 0.05, 0.1, 0.5 and 1
DT	Maximum depth	5:5:100 and None
RF	Number of trees	5:5:100
ANN	Neurons per layer Number of layers Dropout value	2, 3, 4, 5, 7, 10, 15, 20 and 25 1, 2, 3, 4 and 5 10% and None

Table 1: Value of the evaluated hyperparameters

score of 1 indicates flawless detection of all attacks without any false positives. The training time for each model was also analyzed to assess computational efficiency. All experiments were conducted on a system equipped with a 12th Gen Intel(R) Core(TM) i7-1255U processor running at 1.70 GHz.

The outcomes for logistic regression, k-nearest neighbors, and support vector machines are summarized in Tables 2, 3, and 4, respectively. Similarly, the results for decision trees, random forests, and ANNs are presented in Tables 5, 6, and 7. Notably, cross-validation was employed, ensuring that the reported results represent the average performance across the folds. This comprehensive evaluation provides a broad comparison of the selected methodologies under consistent experimental conditions.

Table 2 indicates that logistic regression performs poorly across all configurations tested, with AUC values consistently below 0.6 and an F1-score averaging around 0.46. These results suggest that logistic regression is not a suitable method for the requirements of this case study.

In contrast to the performance of logistic regression, the KNN method achieves excellent results, as highlighted in Table 3. With a small number of neighbors (5 or 10), it delivers AUC and F1-score values exceeding 0.98. Additionally, it maintains a low computational cost, as demonstrated by training times of approximately 0.4 seconds for configurations with lower k values.

Table 4 reveals that the performance of the SVM method varies considerably based on the chosen configuration. The highest results are achieved using a radial

Logistic Regression			
C	AUC	F1-Score	Training Time (s)
0.0001	0.557	0.460	0.058
0.0005	0.556	0.459	0.063
0.001	0.556	0.459	0.065
0.005	0.557	0.460	0.061
0.01	0.557	0.460	0.060
0.05	0.557	0.460	0.060
0.1	0.557	0.460	0.063
0.5	0.557	0.460	0.060
1	0.557	0.460	0.062

Table 2: Logistic Regression results

K-Nearest Neighbors			
k	AUC	F1-Score	Training time (s)
5	0.989769	0.984163	0.404
10	0.987482	0.980886	0.410
25	0.983626	0.970716	0.433
50	0.978763	0.961972	0.462
100	0.965935	0.939058	0.531
250	0.945587	0.908611	0.709
500	0.924719	0.875531	1.017
1000	0.891817	0.827331	1.556

Table 3: K-Nearest Neighbors results

basis function (RBF) kernel with a C value of 1, yielding an AUC of 0.948 and an F1-score of 0.912. However, this method incurs significantly longer training times compared to the other techniques evaluated.

Table 5 shows that decision trees perform exceptionally well across all configurations tested, with AUC and F1-score values consistently close to 1.0. These results indicate that decision trees are highly effective for this case study. Additionally, the training times are relatively short, averaging around 0.13 seconds, making this method both accurate and efficient.

Table 6 demonstrates that the Random Forest method achieves outstanding performance across all configurations tested, with AUC and F1-score values consistently close to 1.0. These results indicate that Random Forest is a highly effective method for this case study. However, the training times increase with the number of trees,

Support Vector Machines				
Kernel	C	AUC	F1-Score	Training time (s)
linear	1	0.574	0.484	55.873
linear	0.1	0.573	0.483	27.710
linear	0.01	0.568	0.476	51.983
poly	1	0.631	0.567	50.607
poly	0.1	0.638	0.572	55.633
poly	0.01	0.675	0.603	41.267
rbf	1	0.948	0.912	21.201
rbf	0.1	0.923	0.876	34.921
rbf	0.01	0.839	0.772	36.838

Table 4: Support Vector Machines results

Decision Trees			
Max Depth	AUC	F1-Score	Training time (s)
5	0.989890	0.984987	0.124396586
25	0.999672	0.999835	0.131734419
50	0.999672	0.999835	0.133901262
75	0.999672	0.999835	0.133561277
100	0.999672	0.999835	0.127586389
None	0.999672	0.999835	0.129640746

Table 5: Decision Trees results

ranging from approximately 0.3 seconds for 5 trees to around 5.2 seconds for 95 trees.

Table 7 shows that the performance of artificial neural networks varies significantly based on the configuration. The best results are achieved with a single layer of 25 neurons and no dropout, yielding an AUC of 0.992 and an F1-score of 0.974. However, the training times are relatively long, averaging around 70 seconds, indicating a higher computational cost than other methods.

Finally, Table 8 summarizes the results obtained using the optimal configurations for each model, trained on the complete training set and evaluated on the remaining samples. Among the methods, DT demonstrates superior performance, achieving an AUC near 0.99 and an F1-score of 0.99, while maintaining a notably short training time.

The outstanding performance of Decision Trees, both in terms of predictive metrics and training efficiency, can be attributed to their ability to capture non-linear

Random Forest			
Number of Trees	AUC	F1-Score	Training time (s)
5	0.99893	0.99846	0.300057
10	0.99930	0.99902	0.584861
15	0.99959	0.99935	0.860937
20	0.99951	0.99927	0.95203
25	0.99946	0.99919	1.115953
45	0.99951	0.99927	1.950927
50	0.99942	0.99911	2.650313
60	0.99946	0.99919	3.424237
65	0.99955	0.99935	3.631376
80	0.99946	0.99919	3.925046
90	0.99942	0.99911	4.209495
100	0.99955	0.99935	4.902763

Table 6: Random Forest results

Artificial Neural Network					
Number of layers	Neurons per layer	Dropout Value	AUC	F1-Score	Training time (s)
1	2	None	0.756511	0.637721	52.039506
1	3	None	0.920820	0.894351	49.520196
1	4	None	0.969060	0.929470	49.040086
1	7	None	0.980585	0.959893	52.675358
1	10	None	0.984818	0.968649	52.769317
1	20	None	0.989573	0.969723	52.675807
1	25	None	0.992563	0.974273	56.451047
2	3	None	0.920605	0.894095	61.865446
2	4	None	0.961834	0.912986	62.665299
2	7	None	0.982982	0.961185	65.715946
2	10	None	0.985042	0.964826	67.794874
2	15	None	0.987199	0.966739	68.610705
2	20	None	0.989318	0.968786	67.688227
2	25	None	0.992282	0.974521	71.208601

Table 7: Artificial Neural Network results

Final Models				
Model	Configuration	AUC	F1-Score	Training time (s)
LR	C=0.0001	0.548	0.445	0.077
KNN	k=5	0.989	0.983	0.511
SVM	Kernel=rbf, C=1	0.950	0.914	47.437
DT	Max.depth=30	0.999	0.999	0.203
RF	Trees=2	0.997	0.997	0.234
ANN	Neurons=25, Layers=2, Dropout=10%	0.996	0.981	93.359

Table 8: Final model results

relationships and handle heterogeneous data without requiring extensive preprocessing [22, 6]. Unlike Logistic Regression, which assumes linear separability, or Support Vector Machines, which often require careful kernel tuning and are computationally intensive, Decision Trees naturally adapt to complex patterns and feature interactions [18, 34]. Compared to KNN, which relies heavily on distance metrics and is sensitive to irrelevant features or scaling, Decision Trees offer a more robust and interpretable structure [22, 7]. While ensemble methods like Random Forests and deep models like Artificial Neural Networks can also achieve high accuracy, they come at the cost of higher computational demand and reduced interpretability [23, 20]. In contrast, Decision Trees provide an effective balance between accuracy, speed, and explainability, which makes them particularly well-suited for the classification task addressed in this study.

6 Conclusions and future works

This work evaluates the effectiveness of six supervised classification methods for detecting DoS attacks in Zigbee networks. It is a comparative analysis using six supervised machine learning algorithms: Logistic Regression, K-Nearest Neighbors, Support Vector Machines, Decision Trees, Random Forest, and Artificial Neural Networks, to identify the most effective system for detecting this type of attack. The evaluation used a real data set consisting of traffic data from a Zigbee network subjected to repeated attacks for model validation.

The results demonstrate the excellent performance of the KNN approach, achieving an AUC value of more than 98% with minimal computational demand. DT and RF also showed outstanding results, with AUC and F1-score values consistently close to 1.0, and relatively short training times. The SVM method obtained commendable results, with AUC values above 94%, but the training time is significantly longer than that of the KNN model. ANN achieved high performance, especially with configurations involving more neurons and layers, but at the cost of longer training times. In contrast, logistic regression proved ineffective for the research objectives.

Regarding real-time deployment, it is important to consider not only the accuracy but also the inference efficiency of each model. While KNN achieved high performance, its computational cost at inference time may limit its use in low-resource edge devices. In contrast, Decision Trees and Random Forests combine high accuracy with low latency and memory usage during inference, making them better suited for scalable, real-time detection systems in diverse IoT environments.

Future research will delve into a broader spectrum of classification algorithms, encompassing supervised, unsupervised, and semi-supervised methods. This exploration aims to uncover more robust and efficient models for detecting DoS attacks in Zigbee networks. Additionally, it is planned to implement these developed models in real-time environments to assess their practical effectiveness and adaptability directly on live data networks. This real-world deployment will provide valuable insights into the models' performance under dynamic conditions and help identify potential areas for further optimization and enhancement.

And the Model-based IDS offer greater flexibility in detecting evolving threats, as they are not limited to predefined rules. Future work will extend the dataset with new attack types like beacon flood and association flood, allowing the models to improve their generalization and robustness against diverse DoS scenarios.

Acknowledgements

This research is the result of the Strategic Project "Critical infrastructures cybersecure through intelligent modeling of attacks, vulnerabilities and increased security of their IoT devices for the water supply sector" (C061/23), as a result of the collaboration agreement signed between the National Institute of Cybersecurity (INCIBE) and the University of A Coruña. This initiative is carried out within the framework of the funds of the Recovery Plan, Transformation and Resilience Plan funds, financed by the European Union (Next Generation).

This work has been funded by the Recovery, Transformation, and Resilience Plan, financed by the European Union (Next Generation) thanks to the "Internet of Things Security in Home and Business Environments in the Context of 5G-IoT Technology" project granted by INCIBE to the University of León.

Míriam Timiraos's research was supported by the "Xunta de Galicia" through grants to industrial PhD (http://gain.xunta.gal/), under the "Doutoramento Industrial 2022" grant with reference: 04_IN606D_2022_2692965.

CITIC, as a Research Center of the University System of Galicia, is funded by Consellería de Educación, Universidade e Formación Profesional of the Xunta de Galicia through the European Regional Development Fund (ERDF) and the Secretaría Xeral de Universidades (Ref. ED431G 2019/01).

Xunta de Galicia. Grants for the consolidation and structuring of competitive research units, GPC (ED431B 2023/49)

References

[1] Mohammed S Alam and Son T Vuong. Random forest classification for detecting android malware. In *2013 IEEE international conference on green computing and communications and IEEE Internet of Things and IEEE cyber, physical and social computing*, pages 663–669. IEEE, 2013.

[2] John Alfred and Ranjit Singh. Enhanced intrusion detection in zigbee networks using machine learning on protocol-specific datasets. *Journal of Wireless Communications and Networks*, 10(5):450–462, 2022.

[3] Abdelouahed Bamou, Moulay Driss EL Ouadghiri, and Badraddine Aghoutane. Current works on ids development strategies for iot. *Lecture Notes in Electrical Engineering*, 745:15–24, 2022.

[4] Mncedisi Bembe, Adnan Abu-Mahfouz, Moshe Masonta, and Tembisa Ngqondi. A survey on low-power wide area networks for iot applications. *Telecommunication Systems*, 71:249–274, 6 2019.

[5] Ekaba Bisong and Ekaba Bisong. Logistic regression. *Building machine learning and deep learning models on google cloud platform: A comprehensive guide for beginners*, pages 243–250, 2019.

[6] Bahzad Charbuty and Adnan Abdulazeez. Classification based on decision tree algorithm for machine learning. *Journal of Applied Science and Technology Trends*, 2(01):20–28, 2021.

[7] Stephan Dreiseitl and Lucila Ohno-Machado. Logistic regression and artificial neural network classification models: a methodology review. *Journal of biomedical informatics*, 35(5-6):352–359, 2002.

[8] Dina Elreedy and Amir F. Atiya. A comprehensive analysis of synthetic minority oversampling technique (smote) for handling class imbalance. *Information Sciences*, 505:32–64, 2019.

[9] Vlad Dacian Gavra, Ovidiu A. Pop, and Ionut Marian Dobra. Performance analysis of xbee modules in a mesh network. *Proceedings of the International Spring Seminar on Electronics Technology*, 2023-May, 2023.

[10] Anand Handa, Ashu Sharma, and Sandeep K Shukla. Machine learning in cybersecurity: A review. *Wiley Interdisciplinary Reviews: Data Mining and Knowledge Discovery*, 9(4):e1306, 2019.

[11] Asma Jahangeer, Sibghat Ullah Bazai, Saad Aslam, Shah Marjan, Muhammad Anas, and Sayed Habibullah Hashemi. A review on the security of iot networks: From network layer's perspective. *IEEE Access*, 11:71073–71087, 2023.

[12] Kamaldeep, Manisha Malik, and Maitreyee Dutta. Feature engineering and machine learning framework for ddos attack detection in the standardized internet of things. *IEEE Internet of Things Journal*, 10:8658–8669, 5 2023.

[13] Ammar Khattak et al. Iot network protocols and their use cases: A comprehensive review. *IEEE Access*, 11:4562–4584, 2023.

[14] Chunyan Li, Jiaji Wang, Shuihua Wang, and Yudong Zhang. A review of iot applications in healthcare. *Neurocomputing*, 565:127017, 1 2024.

[15] Ritika Lohiya and Ankit Thakkar. Application domains, evaluation data sets, and research challenges of iot: A systematic review. *IEEE Internet of Things Journal*, 8:8774–8798, 6 2021.

[16] Knud Lasse Lueth. State of the iot 2020: 12 billion iot connections, 2021.

[17] Alessia Mammone, Marco Turchi, and Nello Cristianini. Support vector machines. *Wiley Interdisciplinary Reviews: Computational Statistics*, 1(3):283–289, 2009.

[18] Álvaro Michelena, José Aveleira-Mata, Esteban Jove, Martín Bayón-Gutiérrez, Paulo Novais, Oscar Fontenla Romero, José Luis Calvo-Rolle, and Héctor Aláiz-Moretón. A novel intelligent approach for man-in-the-middle attacks detection over internet of things environments based on message queuing telemetry transport. *Expert Systems*, page e13263, 2023.

[19] Álvaro Michelena, Francisco Zayas-Gato, Esteban Jove, and José Luis Calvo-Rolle. Detection of dos attacks in an iot environment with mqtt protocol based on intelligent binary classifiers. *Engineering Proceedings*, 7(1):16, 2021.

[20] Rodrigo Moraes, João Francisco Valiati, and Wilson P GaviãO Neto. Document-level sentiment classification: An empirical comparison between svm and ann. *Expert Systems with Applications*, 40(2):621–633, 2013.

[21] Arsalan Mosenia and Niraj K. Jha. Comprehensive survey of security issues in internet of things applications. *IEEE Internet of Things Journal*, 4(6):1753–1770, 2017.

[22] Harsh H Patel and Purvi Prajapati. Study and analysis of decision tree based classification algorithms. *International Journal of Computer Sciences and Engineering*, 6(10):74–78, 2018.

[23] Angshuman Paul, Dipti Prasad Mukherjee, Prasun Das, Abhinandan Gangopadhyay, Appa Rao Chintha, and Saurabh Kundu. Improved random forest for classification. *IEEE Transactions on Image Processing*, 27(8):4012–4024, 2018.

[24] Nasir Raza et al. Security and privacy for iot applications: Protocol analysis. *Journal of Network and Computer Applications*, 106:102941, 2023.

[25] Ankit Sharma and Priya Singh. Analysis of iot protocols for low-power, long-range

networks. *Journal of Communications and Information Networks*, 9(2):125–140, 2023.

[26] Nisha Ashok Somani and Yask Patel. Zigbee: A low power wireless technology for industrial applications. *International Journal of Control Theory and Computer Modelling (IJCTCM)*, 2(3):27–33, 2012.

[27] Björn Stelte and Gabi Dreo Rodosek. Thwarting attacks on zigbee - removal of the killerbee stinger. *2013 9th International Conference on Network and Service Management, CNSM 2013 and its three collocated Workshops - ICQT 2013, SVM 2013 and SETM 2013*, pages 219–226, 2013.

[28] Tole Sutikno and Daniel Thalmann. Insights on the internet of things: past, present, and future directions. *TELKOMNIKA Telecommunication Computing Electronics and Control*, 20(6):1399–1420, 2022.

[29] Ivan Vaccari, M Aiello, and E Cambiaso. Innovative protection system against remote at command attacks on zigbee networks. *Acta Sci. Comput. Sci*, 2:2–8, 2020.

[30] Juanjuan Wang, Mantao Xu, Hui Wang, and Jiwu Zhang. Classification of imbalanced data by using the smote algorithm and locally linear embedding. In *2006 8th international Conference on Signal Processing*, volume 3. IEEE, 2006.

[31] Juanjuan Wang, Mantao Xu, Hui Wang, and Jiwu Zhang. Classification of imbalanced data by using the smote algorithm and locally linear embedding. In *2006 8th international Conference on Signal Processing*, volume 3, 2006.

[32] Liang Wang et al. A comparative study of lpwan protocols for iot applications. *IEEE Internet of Things Journal*, 9(8):6046–6054, 2022.

[33] Wireshark. Wireshark · display filter reference: Zigbee network layer.

[34] Shihong Yue, Ping Li, and Peiyi Hao. Svm classification: Its contents and challenges. *Applied Mathematics-A Journal of Chinese Universities*, 18:332–342, 2003.

[35] Shichao Zhang, Xuelong Li, Ming Zong, Xiaofeng Zhu, and Debo Cheng. Learning k for knn classification. *ACM Transactions on Intelligent Systems and Technology (TIST)*, 8(3):1–19, 2017.

[36] Ángel Luis Muñoz Castañeda, José Antonio Aveleira Mata, and Héctor Aláiz-Moretón. Characterization of threats in iot from an mqtt protocol-oriented dataset. *Complex and Intelligent Systems*, 9:5281–5296, 10 2023.

 Received 12 December 2024

Employing Bayesian Optimization for Hyperparameter Tuning in Convolutional Neural Networks

Amgad M. Mohammed *
Faculty of Information and Communication Technology, Wroclaw University of Science and Technology, Wroclaw, Poland
Faculty of Computers and Information, Menoufia University, Menoufia, Egypt
amgad.elsayed@pwr.edu.pl

Enrique Onieva
DeustoTech, Faculty of Engineering, University of Deusto, Bilbao, Spain
enrique.onieva@deusto.es

Michał Woźniak
Faculty of Information and Communication Technology, Wroclaw University of Science and Technology, Wroclaw, Poland
michal.wozniak@pwr.edu.pl

Abstract

This article focuses on optimizing machine learning when each solution evaluation is costly and/or time-consuming. In particular, current deep learning models, which are complex and computationally intensive, rely on a large number of hyperparameters that need to be optimized. The purpose is to maximize the performance of these models by optimizing their hyperparameters and saving computational resources. In this paper, an informed search using Bayesian optimization is applied to tune the hyperparameters of the Convolutional Neural Network (CNN). Firstly, we analyze the performance of the Stochastic Gradient Descent with Momentum (SGDM) optimizer to learn the weights of CNN faster and more accurately. Then, we show the importance of the momentum hyperparameter for CNN learning. Finally, Bayesian optimization is applied

Special thanks to the anonymous referees
*This work was supported by the Polish National Science Centre under the grant No. 2019/35/B/ST6/04442

to model the generalization performance of CNN as a sample from a Gaussian process (GP). The posterior distribution of the surrogate model is updated sequentially on the basis of a few evaluations of the true expensive function (CNN performance). The results confirm that CNN based on Bayesian optimization has promising performance for hyperparameter optimization when evaluations are expensive or time-limited. We also provide our script code at https://github.com/AmgadMonir/BO-Based-CNN.

Keywords: Hyperparameter optimization; Convolutional neural network; Bayesian optimization; Automated Machine Learning; Stochastic gradient descent; Stochastic gradient descent with momentum; Deep learning optimizers.

1 Introduction

Automated machine learning (AutoML) is known to allow less experienced practitioners to optimally design machine learning (ML) models for a range of tasks while minimizing human error and improving efficiency [18]. AutoML can save significant time and money while improving performance. AutoML includes automation of data collection, automation of data cleansing, automation of feature selection, automation of model selection, hyperparameter optimization, automation of inference, and automating model monitoring and anomaly detection [19]. These topics include hyperparameter optimization (HPO) [4], which improves the reproducibility and fairness of scientific studies better than manual searches for hyperparameters [15]. Furthermore, optimized hyperparameters outperform the default settings offered by typical machine learning libraries [14, 33]. Hyperparameters can be considered as model settings that a data scientist sets before training and controls the capacity of the model. Hyperparameters cannot be learned from the data during the training process. The weights of a deep learning model learned during training are parameters, but the number of hidden layers, the number of neurons, and the type of optimizer are model hyperparameters. Due to its importance, HPO is embedded as part of ML cloud services [5, 7]. HPO is classified as a black-box optimization problem where only the outputs of the ML model can be observed, while the internal dependence on hyperparameters is unknown. The solution to this problem can be reached either by *model-free black box optimization* methods or by *black box Bayesian optimization* methods. The model-free black-box optimization methods include:

- **Manual Search:** The selection of hyperparameters is guided by intuition, experience, or guesswork. We then train the model using these hyperparameters and evaluate its performance on the validation dataset. This process should be repeated until a satisfactory result is obtained.

- **Grid Search:** Is called exhaustive search or a fully factorial design. In grid search, we create a grid with hyperparameter settings, train a model for each combination, and report the evaluation of the validation data. Nevertheless, performance is typically poor for more than a few continuous parameters, and scaling with the dimensionality of the hyperparameter space is expensive. We will never know if we can find the ideal configuration unless we have an infinite hyperparameter lattice that requires infinite time to execute.
- **Random Search:** Randomly samples configurations until a specific search budget is exhausted. This budget can be in the form of a number of function evaluations, time, or money. Random search typically finds a "near optimal" configuration in much fewer iterations compared to grid search [3].
- **Population-based search:** Includes population-based search algorithms [25], such as genetic algorithms, particle swarms, and evolutionary strategies. These algorithms manage a population of configurations and continuously discover the hyperparameter space by switching between exploration and exploitation operators. In addition, these algorithms work well with mixed data types and support parallelization. The covariance matrix adaptation evolution strategy (CMA-ES [17]) is a popular population-based algorithm. Sample configurations from a multivariate Gaussian function and update the mean and covariance in each generation based on the individuals in the population.

In deep learning (DL), we must train multiple networks with different neural architectures, training procedures, and regularization methods and evaluate how well they perform on a validation dataset. Furthermore, which hyperparameters should be adjusted in an algorithm and in what range may be unclear. Finding the ideal hyperparameters is tedious experimentation, even for experts. Population-based search algorithms or evolutionary computing algorithms have been continuously developed for centuries, and numerous standard algorithms are available. Although HPO in DL is particularly challenging when it comes to population-based search algorithms:

- The computation is expensive (an evaluation takes minutes or hours), which precludes a comprehensive evaluation.
- The total number of evaluations is limited by financial, time, or resource constraints.
- The search landscape is multimodal, with different configurations performing similarly.
- Evolutionary algorithms are uninformed methods that do not use past results to decide the next hyperparameters to evaluate.

During optimization, we occasionally need to spend significant resources/time evaluating suboptimal alternatives while looking for the best design. This observation forces us to look for optimization techniques that can intelligently save resources when necessary to find the best settings as quickly as possible [12]. Trying all combinations of hyperparameters is not an option. We can solve all of the above obstacles whenever required using Bayesian optimization (BO). Bayesian optimization allows us to optimize parameters in comparatively few iterations by building a smooth model (called a'surrogate model') from an initial set of parameterizations to predict ML performance for unexplored parameterizations. BO is an adaptive and informed method in which the hyperparameters to be evaluated next are determined from previous evaluations. The same approach can be used to predict the expected performance of all future evaluations and terminate early if the expected gain is less than the observed one. The most important statistical ideas of the last 50 years [13] identified BO as an effective approach.

Unfortunately, the performance of the DL model is susceptible to the choice of hyperparameters. For example, adjusting the learning rate of an optimizer impacts performance. Popular optimizers include Stochastic Gradient Descent (SGD), widely used in machine learning [27]. The aim is to minimize the empirical risk, as explained later. SGD can be written as:

$$\begin{aligned} m^k &= \sum_{i=1}^{k} \beta^{k-i}\delta(i) \\ \omega^{k+1} &= \omega^k - \alpha m^k \end{aligned} \tag{1}$$

where ω denotes the trainable weights of the DL model, $\alpha > 0$ is the learning rate, $\beta \in [0, 1]$ is called the momentum weight. The classical SGD uses $\beta = 0$ and $m^k = \delta(k)$, where $\delta(k)$ is a stochastic gradient of $f(\omega)$ at ω^k. To increase DL performance, a momentum weight of $\beta > 0$ can be used and the resulting algorithm is called SGDM [31]. SGDM uses the exponential moving average of past gradients to avoid getting stuck at saddle points or local minima of non-convex loss surfaces. In this context, SGDM outperforms SGD to minimize the loss function constrained by a tuned value of β. In addition, the momentum parameter β can accelerate convergence when the learning rate α is low. In this article, we demonstrate the convergence rate and the influence of momentum weight on deep neural network learning.

This work aims to optimize CNN performance while accurately classifying different image sets. We cannot estimate the performance of each hyperparameter because training/validation is time-consuming or expensive when renting a cloud service. Instead, Bayesian optimization provides valuable information about which hyperparameter values improve CNN performance better. Therefore, with fewer evaluations, we optimize the hyperparameters better than a comprehensive grid search. The main contributions of this article are as follows.

- Optimizing CNN performance with fewer trials.
- An analysis of the convergence rate for SGD and SGDM.
- An analysis of the effect of the momentum parameter on CNN learning using SGDM.
- Comparison between black-box optimization methods (model-free VS Bayesian optimization).

This article is structured as follows. In Section 2, we introduce the mathematical form of classification based on hyperparameter optimization. Section 3 presents related works on black-box Bayesian optimization and convolutional neural networks. The experimental setup and the experimental results are presented in Section 4. Finally, Section 5 presents the conclusion and the future research line.

2 HPO-based classification

In supervised classification, let $\mathcal{D}$ be a labeled dataset with n input-output pairs $\left(\mathbf{x}^{(i)}, y^{(i)}\right)$, where p-dimensional vector $\mathbf{x}^{(i)} \in \mathcal{X}$ and the label $y^{(i)} \in \mathcal{Y}$. Furthermore, we assume that $\mathcal{D}$ is sampled from an unknown distribution $\mathcal{D} \sim (\mathbb{P}_{xy})^n$. A classification algorithm $\mathcal{I}(\cdot, \lambda)$ configured by hyperparameters $\lambda \in \mathbf{\Lambda}$ maps a dataset $\mathcal{D}$ to a model $f : \mathcal{X} \to \mathcal{Y}$. This model f belongs to the hypothesis space $\mathcal{H} := (\mathbb{D} \times \mathbf{\Lambda})$, where $\mathbb{D}$ is the set of all finite datasets and $\mathbf{\Lambda}$ is the hyperparameter search space with N hyperparameters; $\mathbf{\Lambda} = \Lambda_1 \times \Lambda_2 \times \ldots \Lambda_N$. The domain of a hyperparameter Λ_n can be real-valued (e.g. learning rate), integer-valued (e.g. number of layers) or categorical (e.g. choice of optimizer)[19]. This mixed search space can incorporate dependent hyperparameters λ_i, creating a complicated hierarchical search space. A λ_i is said to be conditional on λ_j if it only becomes active when λ_j is an element of the subset Λ_j; otherwise it remains inactive. As explained in [21], the conditional hierarchy occurs in the optimization of the type of kernel with its parameter values in SVM [9], the automation of machine learning pipelines [6], and the optimization of the neural network architecture [29]. Given a dataset $\mathcal{D}$, the goal is to optimize the expected generalization performance GE of our model based on unseen data $(\mathbf{x}, y) \notin \mathcal{D}$, through the loss function L.

$$\mathrm{GE}(\mathcal{I}, \lambda, \mathcal{D}, L) = \mathbb{E}_{\mathcal{D},(\mathbf{x},y)\sim\mathbb{P}_{xy}} \left[L\left(y, \mathcal{I}_\lambda(\mathcal{D})(\mathbf{x})\right)\right] \tag{2}$$

Given that both $\mathcal{D}$ and a test sample $(\mathbf{x}, y)$ are sampled from $\mathbb{P}_{xy}$. For estimation with multiple random splits, the data $\mathcal{D}$ is split into K sets of $\mathcal{D}^k_{\text{train}}$ and $\mathcal{D}^k_{\text{test}}$ so that $\mathcal{D} = \mathcal{D}^k_{\text{train}} \cup \mathcal{D}^k_{\text{test}}, k = 1, \dots, K$. The expected generalization error can be estimated using equation (3).

$$\widehat{\text{GE}}(\mathcal{I}, \boldsymbol{\lambda}, K, L) := \frac{1}{K} \sum_{k=1}^{K} \frac{1}{|\mathcal{D}^k_{\text{test}}|} \sum_{(\mathbf{x},y)\in\mathcal{D}^k_{\text{test}}} L\left(y, \mathcal{I}_{\boldsymbol{\lambda}}\left(\mathcal{D}^k_{\text{train}}\right)(\mathbf{x})\right) \tag{3}$$

where K is the set of train-test splits, and $\mathcal{I}_{\boldsymbol{\lambda}}(\mathcal{D})$ is given by equation (4) which empirically minimizes inner loss with respect to a point-wise loss function $L(y, f(\mathbf{x})) \to \mathbb{R}$.

$$\mathcal{I}_{\boldsymbol{\lambda}}(\mathcal{D}) = \arg\min_{f\in\mathcal{H}} \mathcal{R}_{\text{emp}}(f), \quad \text{with} \quad \mathcal{R}_{\text{emp}}(f) := \sum_{(\mathbf{x},y)\in\mathcal{D}} L(y, f(\mathbf{x})) \tag{4}$$

From equation (4), the learning algorithm optimizes the empirical risk $\mathcal{R}_{\text{emp}}(f)$ on all candidate models $f \in \mathcal{H}$. The larger the number of hyperparameters Λ_n, the larger the hypothesis space $\mathcal{H}$. In HPO we try to minimize $\widehat{\text{GE}}(\mathcal{I}, \boldsymbol{\lambda}, K, L)$ which is driven mainly by $\boldsymbol{\lambda} \in \boldsymbol{\Lambda}$. Referring to equation 3, HPO is considered a black-box optimization, since the analytical relationship between $\widehat{\text{GE}}(\mathcal{I}, \boldsymbol{\lambda}, K, L)$ and $\boldsymbol{\lambda}$ is unknown [21], and even experts do not know how to efficiently select these hyperparameters. Finally, nested resampling techniques should be used to minimize $\mathcal{R}_{\text{emp}}(f)$ without data leakage from $\mathcal{D}_{\text{test}}$. Consequently, equation (4) can be estimated using an inner resampling technique (e.g., cross-validation) on $\mathcal{D}_{\text{train}}$, while $\mathcal{D}_{\text{test}}$ is used for evaluation. To simplify calculations, especially for large datasets, we can optimize on a single validation by using a simple split: $\mathcal{D} = \mathcal{D}_{\text{train}} \cup \mathcal{D}_{\text{valid}} \cup \mathcal{D}_{\text{test}}$.

An alternate solution can be achieved via the ensemble method [36, 11]. Instead of using the argmin operator to solve equation (4), it might be enough to combine multiple learners $\mathcal{I}_{\boldsymbol{\lambda}}$, each learner is configured by one of the top hyperparameters. As $\mathcal{H}$ becomes large, good configurations can be diverse with the possibility of increasing gain through ensemble methods [10]. To further increase this performance, a stacking model can be learned based on the outputs of the learners formed by HPO [8].

3 Related works

This section introduces the main concepts of Bayesian optimization as an informed search method. The discussion is presented from the perspective of the HPO problem. The second part of this section focuses on convolutional neural networks and their advantages over classical deep neural networks.

3.1 Blackbox Bayesian optimization

Bayesian optimization [30, 35], also called Sequential Model-Based Optimization (SMBO). Bayesian optimization uses a surrogate model that helps machine learning researchers better understand how their black-box functions behave to various hyperparameters. BO typically assumes that the unknown function, the CNN generalization performance in our work, is sampled from a Gaussian process and continuously updates a posterior distribution for that function after each observation $(\lambda \mapsto \mathcal{R}_{\text{emp}}(f))$. The goal is to use all the information available from previous evaluations, $\left[(\lambda^{(1)}, \mathcal{R}_{\text{emp}}(f)^{(1)}), (\lambda^{(2)}, \mathcal{R}_{\text{emp}}(f)^{(2)}), ..., (\lambda^{(i-1)}, \mathcal{R}_{\text{emp}}(f)^{(i-1)})\right]$, rather than relying solely on local gradients and Hessian estimates. This approach minimizes/-maximizes hard nonconvex functions with fewer evaluations but requires additional calculations to determine the next point/hyperparameter to test. The pseudocode of Bayesian optimization is shown in the algorithm 1. BO showed good results in optimizing deep learning models for image classification [32], speech recognition [37], and text generation [20]. Next, we introduce the two main parts of Bayesian optimization; the probabilistic surrogate model and the acquisition function.

Algorithm 1: The pseudocode of Bayesian Optimization Algorithm.

```
Input: Initial dataset D
repeat
    λ ← POLICY (D) ;                     ▷ Optimize acquisition function.
    R_emp(f) ← OBSERVE (λ) ;             ▷ Observe at the selected λ.
    D ← D ∪ (λ, R_emp(f)) ;              ▷ Update the dataset.
until termination condition is met ▷ exhausted budget;
```

3.1.1 Surrogate model

The Gaussian process (GP) is a non-parametric statistical tool for solving regression problems and statistical classification tasks. GP can be viewed as a distribution over functions, where each function value follows a Gaussian (normal) distribution at any point in time. A GP can be written as:

$$\mathcal{R}_{\text{emp}}(f) \sim \mathcal{GP}\left(\mu(\lambda), k\left(\lambda^1, \lambda^2\right)\right)$$

where $\mathcal{R}_{\text{emp}}(f)$ is a function sampled from a Gaussian process, and $\mu(\lambda)$ and $k\left(\lambda^1, \lambda^2\right)$ are the mean and covariance functions, respectively. The mean function $\mu(\lambda)$ expects the value of the function at any given point λ. The smoothness of the function learned by GP is influenced by the covariance matrix. The covariance function or kernel, $k\left(\lambda^1, \lambda^2\right)$, defines the relationship between points in the input space. The closer the points are, the higher their correlation. This codes the belief that they have similar function values different from points that are far apart. GP's model has the following advantages: (1) The prediction interpolates the observations. (2) The prediction is probabilistic (Gaussian) in nature, allowing the calculation of empirical confidence intervals. (3) Flexibility in modeling continuous functions with unknown properties. However, GP is not sparse, but considers all features to make prediction. This makes GP less efficient in high-dimensional spaces, dozens of hyperparameters. Furthermore, performance depends heavily on the choice of the kernel function and its hyperparameters.

3.1.2 Acquisition function

An acquisition function is created by combining the predictive mean function $\mu\Big(\lambda; \{\lambda^{(n)}, \mathcal{R}_{\text{emp}}(f)^{(n)}\Big)$, and the predictive variance function, $\sigma^2(\lambda; \{\lambda^{(n)}, \mathcal{R}_{\text{emp}}(f)^{(n)})$, from the surrogate model for previous observations. This function reflects the importance of observing a particular parameterization. We optimize the acquisition function to determine which hyperparameter is optimal for the next observation $\lambda^{(i+1)}$. Once we observe the performance of the ML in that configuration, $\mathcal{R}_{\text{emp}}(f)^{(i+1)}$, we fit a new surrogate model and continue the procedure until convergence. The predictions and uncertainty estimates are revised sequentially in response to iterative observations/evaluations.

Expected Improvement (EI) is a common acquisition function due to its easy-to-calculate analytical form and excellent practical performance. It awards an evaluation of the target function based on expected improvement beyond the current best $\mathcal{R}_{\text{emp}}(f)^*$. EI is defined in equation (5), where $h(\lambda)^{(i+1)}$ is the posterior mean of the surrogate at the $(i+1)^{th}$ iteration. The parameterization λ with the highest EI is selected and evaluated. At the beginning of the search, EI explores regions where the uncertainty measurement is high and, after a few iterations, exploits regions with a high probability of a good objective value. Figure 1 shows seven iterations of EI during the maximization of a ground truth function [1]. With more iterations, the uncertainty about the true function decreases, and it becomes easier to locate hyperparameter values where the function peaks.

$$\text{EI }(\lambda) = \text{E}\left[\max\Big(h(\lambda)^{(i+1)} - \mathcal{R}_{\text{emp}}(f)^*, 0\Big)\right] \tag{5}$$

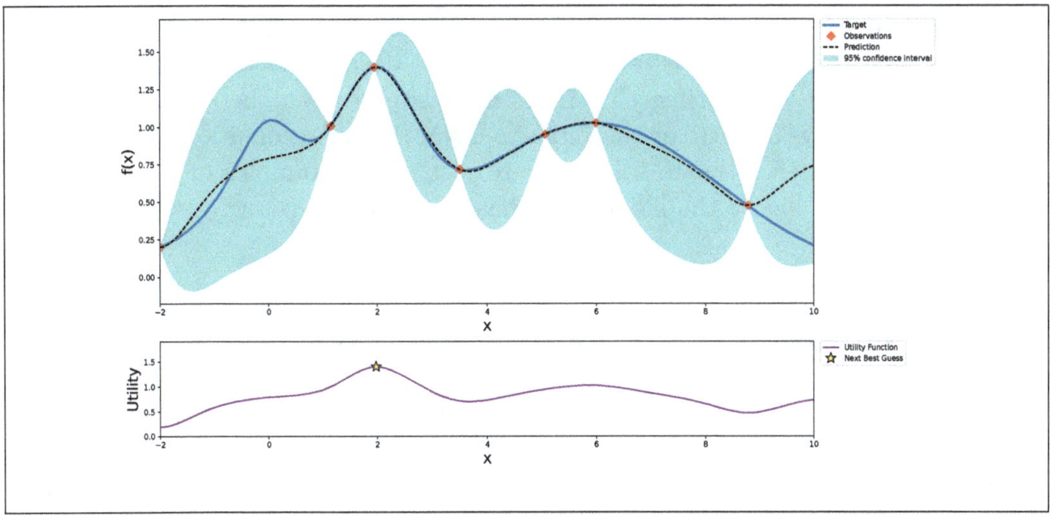

Figure 1: Bayesian Optimization with Gaussian Process and Expected Improvement.

3.2 Convolutional neural networks

The biggest challenge in deep neural networks is overfitting. Overfitting is caused by having more parameters than necessary and limited data points to learn those weights. A well-known and widely used mechanism to avoid overfitting is a structure called *convolution.* Convolution allows weights to be reused to detect the same pattern in multiple places. Surprisingly, this technique makes our model more accurate and smaller in terms of the number of weights to learn and store [34]. The four main features and components of CNN, as described in [26], are: (1) local connections: Instead of a large dense layer of connections going from every input to every output, we have a *Convolutional layer* consisting of many convolution kernels. The convolution kernel is a small matrix that slides over the entire image to produce feature maps such as horizontal edges, vertical edges, and even complex features. (2) Padding: ensures the kernel fits appropriately over the image, especially around the edges. Without padding, each convolution operation results in a smaller feature map. (3) Stride: the step size of the filter to traverse the input image. If the step is set to 1, the filter moves vertically and horizontally, one pixel at a time. A larger step reduces the size of the feature map as the filter bypasses more pixels. (4) Pooling layers: are downsampling dimensionality reduction techniques that help reduce the amount of data while retaining useful information and accelerating convergence. CNNs [24, 23] are ideal for processing image data and natural language because they automatically detect spatial hierarchies, including textures and complex patterns, as long as we add deep layers. In the ImageNet Large-Scale Visual Recognition Challenge (LSVRC), Krizhevsky et al. [23] used deep CNN to achieve the best classification result at the time. Their results inspired researchers and developers to advance the development of the modern CNN. Facial recognition, autonomous vehicles, and smart healthcare solutions are just a few examples of things people have been able to achieve thanks to CNN-based computer vision.

3.3 BO-Based CNN

In [2], BO is coupled with CNN for medical image classification. In particular, to automate the selection of the most suitable hyperparameters of CNN for the purpose of classifying different types of brain tumors. In their article, BO is run with 40 iterations to optimize the following set of hyperparameters (type of activation function, batch size, dropout rate, number of dense nodes, type of optimizer). The limitations of their proposal are (1) HPO is not explained mathematically as a type of black-box optimization. (2) A shallow discussion of Grid search and Random search is mentioned in their work, with no evidence about their limited capabilities against BO in their results section. (3) There is no script code to facilitate the reproducibility of their findings. In [16], BO is integrated with CNN for wind power forecasting. The authors employed BO to optimize the sampling among the following set of hyperparameters (number of kernels of each convolutional layer, number of fully connected layers, type of activation function, type of loss function, type of optimizer, number of epochs). The advantages of their work are (1) the authors analyzed the performance of several frameworks that support implementation of BO: Scikit optimize [1], Optuna [2] and Hyperopt [3]. (2) The top 10 combinations of hyperparameters are analyzed more deeply with different seeds to select the best CNN architecture that is less sensitive to randomness. However, the limitations of their work are still the same as [2]. Similarly in [22], BO-CNN is used for machine fault diagnosis. The authors categorized CNN hyperparameters into network structure definition hyperparameters and network training process hyperparameters. By altering the number of convolutional layers, the number of kernels, and the size of the kernel, the optimization algorithm automates the selection of the best CNN structure. In addition, network training hyper-parameters; *learning rate, momentum* can be optimized for better performance. To this end, our article contributes to solve the limitations aforementioned from the previous work while considering a fixed CNN structure with a focus on optimizing training process hyperparameters

[1]Scikit optimize: https://scikit-optimize.github.io/stable/
[2]Optuna: https://optuna.org/
[3]Hyperopt: https://github.com/hyperopt/hyperopt

(*learning rate, momentum*).

4 Experimental results

We use CNN to classify according to MNIST and CIFAR10 from PyTorch datasets and apply Bayesian optimization to optimize the SGDM settings for the best CNN performance.

- $\alpha \rightarrow$ The learning rate hyperparameter determines the step size used in gradient descent in the neural network. It takes real values in the range $[10^{-6}, 1]$.
- $\beta \rightarrow$ The momentum weight hyperparameter determines the percentage contribution of past gradients. It takes real values in the range $[0, 1]$.

To simplify the calculations, we perform a three-part breakdown; $\mathcal{D} = \mathcal{D}_{\text{train}} \cup \mathcal{D}_{\text{valid}} \cup \mathcal{D}_{\text{test}}$, The original training data is divided into 80% for training, 20% for validation, and the test set is invisible. Matern 5/2 kernel for the GP model is employed as one of the most frequently used and flexible choices for BO. We used the Facebook-supported AX [4] framework as open source for the adaptive experiment and BO implementation. The loss of cross-entropy [28] is the function to be optimized to train the CNN model, without early stopping and no learning rate decay strategies. Figure 2 shows the convolutional neural network architectures for the classification of the MNIST and CIFAR10 datasets. For MNIST classification, the CNN contains an input, a single block (one convolutional layer and one max-pooling layer) and a classification block (two fully connected layers). A convolution layer that employs 20 2-D kernels, each 5×5 in size, is applied to every image with the inclusion of zero padding. Activation function (ReLU) is applied to the output of the convolution layer which is 20@24×24, followed by a reduction through the max-pooling layer with 3×3 kernel to produce 20 feature maps, each 8×8 in size. Finally, the classification block has two fully connected layers, each containing 64 and 10 neurons, respectively. For CIFAR10 classification, the CNN architecture is composed of three main blocks (three convolutional and three max-pooling layers), and a classification block (three fully connected layers). All convolutional laryers apply kernels of size 3×3 with the inclusion of padding = 1 to the entry image. The only difference is with the number of kernels in each convolutional layer being 32, 32, and 64 sequentially. Activation function (ReLU) is applied, followed by a max-pooling layer with 2×2 kernel size. Batch normalization is applied after each main block, after the max-pooling layer, to normalize the outputs from the inner layers. Finally, the classification block has three fully connected layers, each containing 512, 64 and 10 neurons, respectively.

[4] AX Framework: https://ax.dev/docs/tutorials/quickstart/

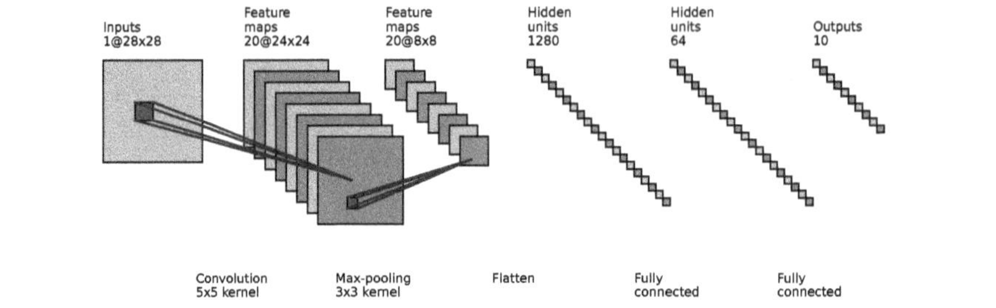

(a) CNN Architecture For MNIST Classification.

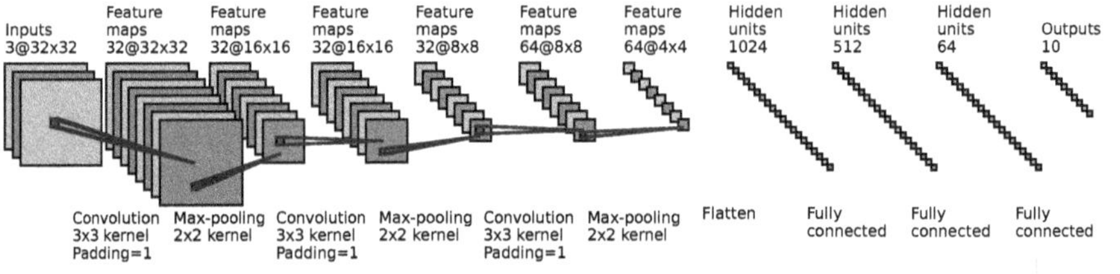

(b) CNN Architecture For CIFAR10 Classification.

Figure 2: Different CNN architectures based on the complexity of classification task.

The CNN architectures in Figure 2 are generated using the draw convnet code [5].

4.1 Analysis of convergence rate

The convergence rate describes how quickly an optimizer approaches the minimum of the loss function from the starting points. This is particularly helpful in determining when to stop numerical calculations, such as after a certain level of precision has been achieved. Let $\epsilon^k = |\omega^k - \omega^*|$ denote the error, where ω^* denotes the optimal value. If we do not know ω^*, we stop the algorithm after N iterations, and the error becomes $\epsilon^k = |\omega^k - \omega^N|$. The optimizer converges when $\lim\limits_{k\to\infty} \epsilon^k = 0$. We are interested in calculating the ratio $\epsilon^{k+1}/\epsilon^k$. The convergence of ϵ has an order q if $\lim\limits_{k\to\infty} \epsilon^{k+1}/(\epsilon^k)^q < C$ consistent with one of the following:

- $q = 1$ and $C \in (0, 1) \rightarrow$ linear convergence.
- $q = 2$ and $C > 0 \rightarrow$ quadratic convergence.
- $q = 1$ and $C = 1 \rightarrow$ sublinear convergence (slower than linear).

[5]draw convnet https://github.com/gwding/draw_convnet

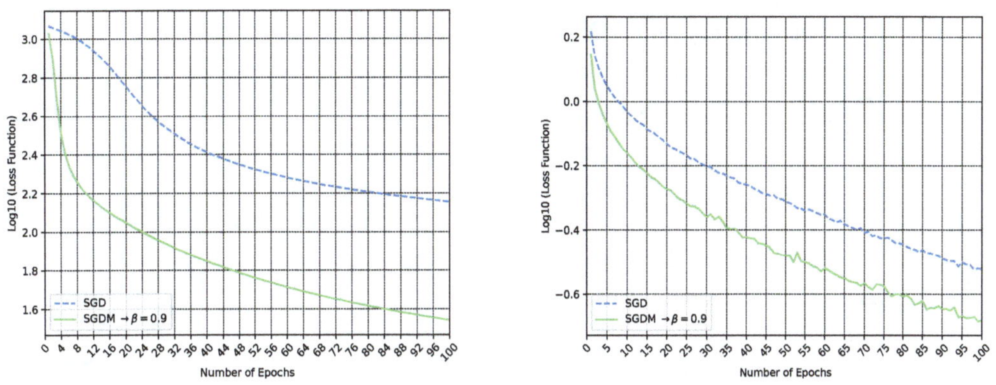

(a) Performance on MNIST with $\alpha = 1e-6$ (b) Performance on CIFAR10 with $\alpha = .001$

Figure 3: Convergence rate analysis for SGD and SGDM.

When applying classical SGD, the gradient of the loss function at saddle points (plateau) is negligible or zero, which in turn results in little or no weight updates. Even after many iterations, the loss function is reduced very slowly. In contrast, in SGDM, the loss function gradient at the saddle points is influenced by the exponential moving average of past gradients, which causes the weight update to be large and to surpass the saddle point. Figure 3 shows the minimization of the loss function during CNN training. The left side of the figure 3 confirms that the minimum value at 80th epoch by SGD can be easily reached after 10 epochs by SGDM. Similarly, the right side of figure 3 shows that SGDM in the 35th epoch is comparable to SGD in the 70th epoch. Both SGD and SGDM have a sublinear convergence rate, but SGDM can accelerate CNN learning, as our results confirm. For more information on algorithms' convergence analysis, see [6].

4.2 Impact of momentum weight on learning

This part of the experiments analyzes the effect of the momentum weight β. From equation (1), it is clear that small β reduces the contribution of previous gradients. Setting $\beta = 0$ activates the classic SGD in which the gradients disappear at local minima and the weight learning stops. However, larger values of β take into account more gradients from the past. Figure 4 shows the average validation accuracy for 10 SGDM runs with different settings, where α is fixed, and $\beta \in \{0.4, 0.45, ..., 0.9, 0.95\}$. Furthermore, the average of 10 runs of classic SGD is significantly suppressed by

[6]Scientific Computing with Python: https://caam37830.github.io/book

each SGDM setting, as shown for both datasets. The results are reported for each configuration after training our CNN with 20 epochs.

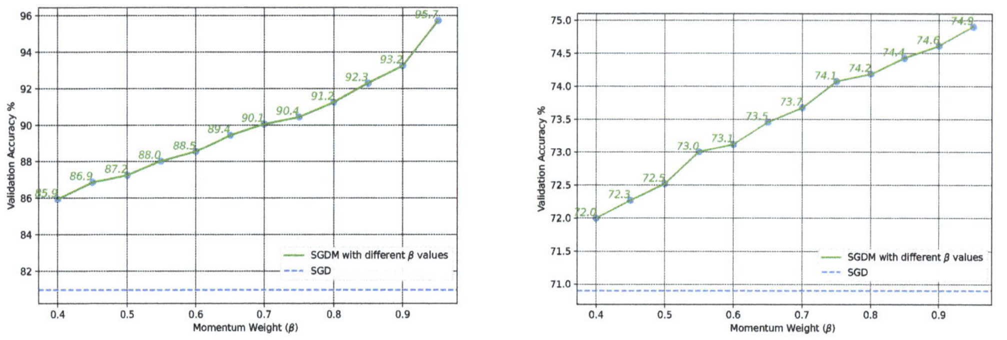

(a) Performance on MNIST with $\alpha = 1e-6$

(b) Performance on CIFAR10 with $\alpha = .001$

Figure 4: Impact of momentum weight on CNN learning.

4.3 Comparison of blackbox optimization methods

The actual impact of each β on learning is further influenced by the value of α. Therefore, both hyperparameters should be adjusted at the same time. We will not present the global optimum for actual CNN performance as this is too expensive. Rather, it may be sufficient to compare different black-box optimization methods in Figure 5. We ran a random search three times with different seeds and considered the mean at each iteration/trial. For each iteration, the maximum number of epochs was 20. Figure 5 proves the usefulness of hyperparameter optimization, shows the maximum validation accuracy that is reached up to the current evaluation number. With only 30 evaluations/attempts, Bayesian optimization can find a good hyperparameter combination, $\lambda = [\alpha, \beta]$, with the highest validation accuracy on both datasets. Furthermore, we clarify that random search is comparable to Bayesian optimization for the following reasons: (1) the low dimensionality of the search space, (2) in our implementation, we have defined the hyperparameter grid from which random search can select its random value; otherwise, we notice a big degradation in its performance. The initial sub-par performance of Bayesian optimization can be attributed to the initial exploration. In addition, the grid search strategy requires multiple iterations to be competitive, as we show in the next section.

Due to the good performance of Bayesian optimization to discover the search space. Figure 6 shows the contour plots of the modeled CNN performance as a function of the hyperparameters α and β. The heuristic search of the acquisition

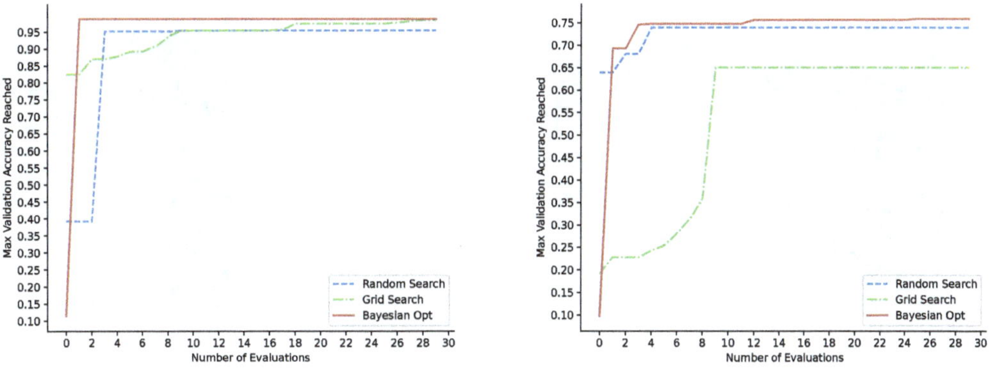

(a) Performance on MNIST.

(b) Performance on CIFAR10.

Figure 5: Model-Free Blackbox Optimization Versus Blackbox Bayesian Optimization.

function allows the ML practitioner to select appropriate configurations, most of them with higher CNN performance.

4.4 Validation plots and test evaluation.

The validation plots in Figure 7 confirm the accuracy with which the surrogate model (GP) simulates the actual performance of CNN. If all points are distributed around the dashed line, the model is a good predictor of the outcome of the expensive function. Each subgraph contains information from the thirty evaluations. The scale of y-axis reflects how easier is the classification task, MNIST is much easier than CIFAR10.

Finally, table 1 shows the optimized hyperparameters, the number of iterations of the best configurations, the validation accuracy and the testing accuracy. To report the test results, the training and validation sets are combined. After that, CNN is trained on the combined set with the best configuration found by each search method, followed by an evaluation of the test set. Among the three random search runs, we report the results of the best. The grid search achieves the best test accuracy at the expense of multiple evaluations of the expensive function. In particular, the grid search reaches the optimized hyperparameters at iterations of 39th and 72th for both MNIST and CIFAR10 respectively. In contrast, Bayesian optimization achieves competitive test results, with significantly fewer iterations, at iterations of 2nd and 26th for both MNIST and CIFAR10. Again, the performance of both random search and grid search can only be achieved through our well-defined

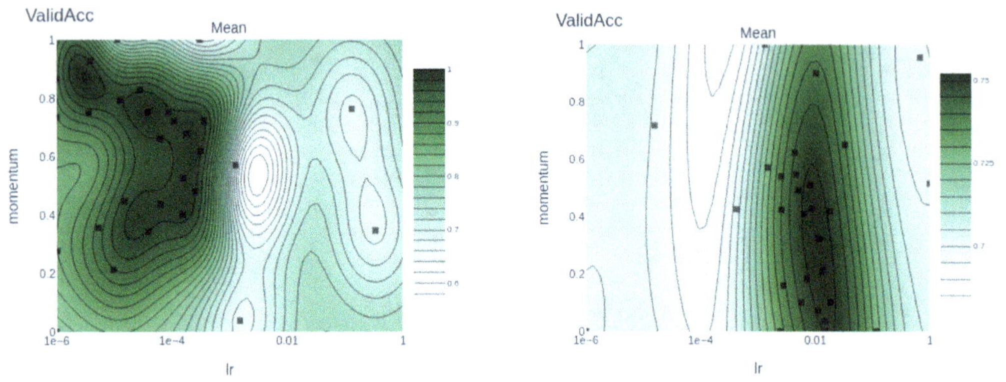

(a) Performance on MNIST. (b) Performance on CIFAR10.

Figure 6: Contour Plots of the Modeled CNN Performance.

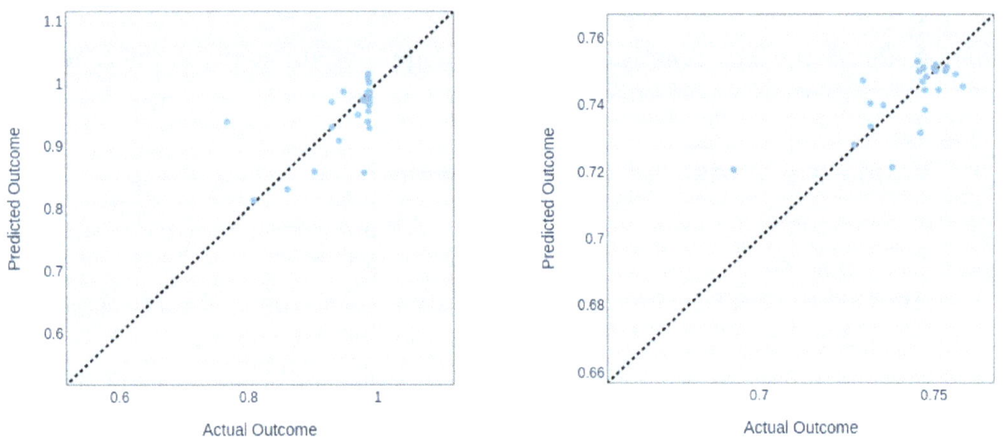

(a) Performance on MNIST. (b) Performance on CIFAR10.

Figure 7: Validation Plots of Surrogate Models Against Real CNN Performance.

hyperparameter search space in advance, while BO requires only hyperparameter bounds.

5 Conclusion

This article examined Bayesian Optimization as a method for optimizing a black-box function (with unknown gradients). It is especially advantageous when the function

Dataset	α, β, evaluation metrics	Random Search	Grid Search	Bayesian Optimization
MNIST	Learning rate (α)	0.00031	0.00007	0.00036
	Momentum weight (β)	0.77778	0.88889	0.72202
	Best iteration	14	39	**2**
	Validation Accuracy	**98.83**	98.82	98.78
	Test Accuracy	98.99	**99.02**	98.94
CIFAR10	Learning rate (α)	0.00543	0.02276	0.01832
	Momentum weight (β)	0.77778	0.11111	0.41898
	Best iteration	**2**	72	26
	Validation Accuracy	75.54	75.81	**75.83**
	Test Accuracy	76.87	**77.43**	76.16

Table 1: The optimized hyperparameters, best iteration number, validation and test accuracies.

evaluation is expensive and requires a lot of computation or even some interaction with the outside world. In those cases, population-based search and grid/exhaustive search methods are unfeasible solvers. We discussed the fundamental components of Bayesian optimization. The specific application was to tune the hyperparameters of convolutional neural networks with fewer trials intelligently, to match the budget associated with training the model. Initially, we analyze CNN performance on the basis of optimizer type. In particular, stochastic gradient descent with momentum (SGDM) was highly efficient and faster in minimizing the CNN training loss function. We then analyzed the influence of momentum weight on validation accuracy. The second part of this article focuses on optimizing SCDM settings and DL model hyperparameters using various search techniques such as random search, grid search, and Bayesian optimization. The sequential evaluations of CNN configured by different hyperparameters are used as data by the Gaussian process to obtain the surrogate posterior. Subsequently, an acquisition function can be optimized to find promising hyperparameters. The results confirm the superiority of Bayesian optimization in improving CNN performance without a trial/error mechanism, thereby contributing to the field of automated machine learning. The effectiveness of Bayesian optimization depends on the ability of the proxy to model the actual black-box function. This is why we show the validation diagrams. Furthermore, random search is significantly more effective than grid search in exploring the search space, given an equal number of iterations.

The framework can be extended for future research by applying constrained Bayesian optimization (CBO), i.e., optimizing the hyperparameters of deep learning models under resource constraints; like memory consumption, training time, prediction time, and energy consumption. In particular, tuning the number of layers affects accuracy, memory, and energy. Furthermore, tuning the learning rate affects the accuracy and convergence time. To tune deep learning models while respecting system or deployment limits; the set of constraints could involve training time ≤ 2 hours and/or memory usage ≤ 4 GB. Then, modified acquisition functions can be employed to find feasible hyperparameters. CBO is particularly valuable where performance must be balanced against limited resources; like embedded systems, edge computing, cloud model deployment, healthcare AI, robotics, gaming, and more. In addition, each constraint can be treated as a single black-box function and multiobjective Bayesian optimization can be applied for better trade-offs between performance and cost. Additionally, multi-fidelity optimization techniques can speed up evaluating expensive black-box functions, resulting in faster Bayesian optimization.

References

[1] Apoorv Agnihotri and Nipun Batra. Exploring bayesian optimization. *Distill*, 5(5):e26, 2020.

[2] Mohamed Ait Amou, Kewen Xia, Souha Kamhi, and Mohamed Mouhafid. A novel mri diagnosis method for brain tumor classification based on cnn and bayesian optimization. In *Healthcare*, volume 10, page 494. MDPI, 2022.

[3] James Bergstra and Yoshua Bengio. Random search for hyper-parameter optimization. *Journal of machine learning research*, 13(2), 2012.

[4] Bernd Bischl, Martin Binder, Michel Lang, Tobias Pielok, Jakob Richter, Stefan Coors, Janek Thomas, Theresa Ullmann, Marc Becker, Anne-Laure Boulesteix, et al. Hyperparameter optimization: Foundations, algorithms, best practices, and open challenges. *Wiley Interdisciplinary Reviews: Data Mining and Knowledge Discovery*, 13(2):e1484, 2023.

[5] Ekaba Bisong and Ekaba Bisong. Google automl: cloud vision. *Building Machine Learning and Deep Learning Models on Google Cloud Platform: A Comprehensive Guide for Beginners*, pages 581–598, 2019.

[6] Zoltan Czako, Gheorghe Sebestyen, and Anca Hangan. Automaticai–a hybrid approach for automatic artificial intelligence algorithm selection and hyperparameter tuning. *Expert Systems with Applications*, 182:115225, 2021.

[7] Piali Das, Nikita Ivkin, Tanya Bansal, Laurence Rouesnel, Philip Gautier, Zohar Karnin, Leo Dirac, Lakshmi Ramakrishnan, Andre Perunicic, Iaroslav Shcherbatyi, et al. Amazon sagemaker autopilot: a white box automl solution at scale. In *Proceedings of the fourth international workshop on data management for end-to-end machine learning*, pages 1–7, 2020.

[8] Alfredo Daza, Juana Bobadilla, Juan Carlos Herrera, Angelica Medina, Nemias Saboya, Karoline Zavaleta, and Segundo Siguenas. Stacking ensemble based hyperparameters to diagnosing of heart disease: Future works. *Results in Engineering*, 21:101894, 2024.

[9] Laura Dioşan, Alexandrina Rogozan, and Jean-Pierre Pecuchet. Improving classification performance of support vector machine by genetically optimising kernel shape and

hyper-parameters. *Applied Intelligence*, 36:280–294, 2012.

[10] Matthias Feurer, Aaron Klein, Katharina Eggensperger, Jost Springenberg, Manuel Blum, and Frank Hutter. Efficient and robust automated machine learning. *Advances in neural information processing systems*, 28, 2015.

[11] Mudasir A Ganaie, Minghui Hu, Ashwani Kumar Malik, Muhammad Tanveer, and Ponnuthurai N Suganthan. Ensemble deep learning: A review. *Engineering Applications of Artificial Intelligence*, 115:105151, 2022.

[12] Roman Garnett. *Bayesian optimization.* Cambridge University Press, 2023.

[13] Andrew Gelman and Aki Vehtari. What are the most important statistical ideas of the past 50 years? *Journal of the American Statistical Association*, 116(536):2087–2097, 2021.

[14] Joe G Greener, Shaun M Kandathil, Lewis Moffat, and David T Jones. A guide to machine learning for biologists. *Nature reviews Molecular cell biology*, 23(1):40–55, 2022.

[15] Odd Erik Gundersen, Kevin Coakley, Christine Kirkpatrick, and Yolanda Gil. Sources of irreproducibility in machine learning: A review. *arXiv preprint arXiv:2204.07610*, 2022.

[16] Shahram Hanifi, Andrea Cammarono, and Hossein Zare-Behtash. Advanced hyperparameter optimization of deep learning models for wind power prediction. *Renewable Energy*, 221:119700, 2024.

[17] Nikolaus Hansen. The cma evolution strategy: A tutorial. *arXiv preprint arXiv:1604.00772*, 2016.

[18] Xin He, Kaiyong Zhao, and Xiaowen Chu. Automl: A survey of the state-of-the-art. *Knowledge-based systems*, 212:106622, 2021.

[19] Frank Hutter, Lars Kotthoff, and Joaquin Vanschoren. *Automated machine learning: methods, systems, challenges.* Springer Nature, 2019.

[20] Deepali Jain, Malaya Dutta Borah, and Anupam Biswas. Fine-tuning textrank for legal document summarization: A bayesian optimization based approach. In *Proceedings of the 12th Annual Meeting of the Forum for Information Retrieval Evaluation*, pages 41–48, 2020.

[21] Florian Karl, Tobias Pielok, Julia Moosbauer, Florian Pfisterer, Stefan Coors, Martin Binder, Lennart Schneider, Janek Thomas, Jakob Richter, Michel Lang, et al. Multi-objective hyperparameter optimization in machine learning—an overview. *ACM Transactions on Evolutionary Learning and Optimization*, 3(4):1–50, 2023.

[22] Davor Kolar, Dragutin Lisjak, Michał Pająk, and Mihael Gudlin. Intelligent fault diagnosis of rotary machinery by convolutional neural network with automatic hyper-parameters tuning using bayesian optimization. *Sensors*, 21(7):2411, 2021.

[23] Alex Krizhevsky, Ilya Sutskever, and Geoffrey E Hinton. Imagenet classification with deep convolutional neural networks. *Advances in neural information processing systems*, 25, 2012.

[24] Yann LeCun, Léon Bottou, Yoshua Bengio, and Patrick Haffner. Gradient-based learn-

ing applied to document recognition. *Proceedings of the IEEE*, 86(11):2278–2324, 1998.

[25] Wendi Li, Ting Wang, and Wing WY Ng. Population-based hyperparameter tuning with multitask collaboration. *IEEE transactions on neural networks and learning systems*, 34(9):5719–5731, 2021.

[26] Zewen Li, Fan Liu, Wenjie Yang, Shouheng Peng, and Jun Zhou. A survey of convolutional neural networks: analysis, applications, and prospects. *IEEE transactions on neural networks and learning systems*, 33(12):6999–7019, 2021.

[27] Zijian Liu, Ta Duy Nguyen, Thien Hang Nguyen, Alina Ene, and Huy Nguyen. High probability convergence of stochastic gradient methods. In *International Conference on Machine Learning*, pages 21884–21914. PMLR, 2023.

[28] Anqi Mao, Mehryar Mohri, and Yutao Zhong. Cross-entropy loss functions: Theoretical analysis and applications. In *International conference on Machine learning*, pages 23803–23828. PMLR, 2023.

[29] Pengzhen Ren, Yun Xiao, Xiaojun Chang, Po-Yao Huang, Zhihui Li, Xiaojiang Chen, and Xin Wang. A comprehensive survey of neural architecture search: Challenges and solutions. *ACM Computing Surveys (CSUR)*, 54(4):1–34, 2021.

[30] Bobak Shahriari, Kevin Swersky, Ziyu Wang, Ryan P Adams, and Nando De Freitas. Taking the human out of the loop: A review of bayesian optimization. *Proceedings of the IEEE*, 104(1):148–175, 2015.

[31] Bin Shi. On the hyperparameters in stochastic gradient descent with momentum. *Journal of Machine Learning Research*, 25(236):1–40, 2024.

[32] Jasper Snoek, Oren Rippel, Kevin Swersky, Ryan Kiros, Nadathur Satish, Narayanan Sundaram, Mostofa Patwary, Mr Prabhat, and Ryan Adams. Scalable bayesian optimization using deep neural networks. In *International conference on machine learning*, pages 2171–2180. PMLR, 2015.

[33] Chris Thornton, Frank Hutter, Holger H Hoos, and Kevin Leyton-Brown. Auto-weka: Combined selection and hyperparameter optimization of classification algorithms. In *Proceedings of the 19th ACM SIGKDD international conference on Knowledge discovery and data mining*, pages 847–855, 2013.

[34] Andrew W Trask. *Grokking deep learning.* Simon and Schuster, 2019.

[35] Xilu Wang, Yaochu Jin, Sebastian Schmitt, and Markus Olhofer. Recent advances in bayesian optimization. *ACM Computing Surveys*, 55(13s):1–36, 2023.

[36] Florian Wenzel, Jasper Snoek, Dustin Tran, and Rodolphe Jenatton. Hyperparameter ensembles for robustness and uncertainty quantification. *Advances in Neural Information Processing Systems*, 33:6514–6527, 2020.

[37] Boyang Xue, Shoukang Hu, Junhao Xu, Mengzhe Geng, Xunying Liu, and Helen Meng. Bayesian neural network language modeling for speech recognition. *IEEE/ACM Transactions on Audio, Speech, and Language Processing*, 30:2900–2917, 2022.

 Received 15 December 2024

Noise-free Wavefront Prediction for Open-loop Adaptive Optics

Saúl Pérez, Carlos González-Gutiérrez
Instituto de Ciencias y Tecnologías Espaciales de Asturias (ICTEA), University of Oviedo, Oviedo, Spain and Department of Computer Science, University of Oviedo, Oviedo, Spain

Alejandro Buendía
Instituto de Ciencias y Tecnologías Espaciales de Asturias (ICTEA), University of Oviedo, Oviedo, Spain and Department of Mathematics, University of Oviedo, Oviedo,33007, Spain

Santiago Iglesias, Javier Rodríguez, Julia Fernández
Instituto de Ciencias y Tecnologías Espaciales de Asturias (ICTEA), University of Oviedo, Oviedo, Spain

Francisco Javier de Cos
Instituto de Ciencias y Tecnologías Espaciales de Asturias (ICTEA), University of Oviedo, Oviedo, Spain and Department of Exploitation and Exploration of Mines, University of Oviedo, Oviedo, Spain

Abstract

Ground-based telescopes face a significant challenge posed by atmospheric turbulence, which causes images to appear blurred and lack clarity. Adaptive optics technology is employed to mitigate this issue by effectively correcting wavefront aberrations through adjustments to the surface of a deformable mirror. Moreover, incorporating neural networks into the control system has shown significant improvements in both atmospheric correction and turbulence prediction. Specifically, in this work, a 2D-LSTM network structure is utilized, which has shown good efficiency in slope prediction over a time sequence. The aim of this study is to forecast turbulence data while removing noise introduced by reading instruments. Through experiments conducted in this research, utilizing a dataset of 40,000 clean and 40,000 noisy instances simulated with Soapy, it is demonstrated that these neural models are capable of learning noise patterns, resulting in output data that closely mimic actual turbulence conditions.

Keywords: Adaptive Optics Machine Learning Prediction.

1 Introduction

Atmospheric turbulence acts like a lens that continuously modifies the refractive index of the layers composing the atmosphere. This results in distortions in the wavefront, which directly lead to obtaining images with lower quality than desired. This degrades useful information about the observed object and may introduce errors in associated calculations for numerous studies and experiments. To address this problem, a technology known as Adaptive Optics (AO) is used. Two main phases can be distinguished in the operation of an AO system. First, a wavefront sensor, typically a Shack-Hartmann (SHWFS), decomposes the incoming wavefront using a microlens array, focusing small portions of the wavefront onto a detector. The relative lens shifts in the newly generated focal spots will allow one to compute their centroids, from which the slopes of the wavefront can be acquired, showing the local tilts of the wavefront. This information will be passed on to the control system, triggering the second operational phase: wavefront reconstruction and correction based on acquired measurements. Essentially, a deformable mirror with a finite number of actuators is activated to modify its surface, allowing the reflected light to reconstruct the original wavefront as accurately as possible, yet with irreducible residual error remaining.

Within Adaptive Optics (AO) systems, a wide variety of configurations can be found. However, this study will focus on a Single-Conjugated AO (SCAO) system in open loop. This type of system uses a single deformable mirror and corrects aberrations based on the measurements from a single wavefront sensor, optimizing the correction around a nearby bright guide star.

In recent years, several wavefront reconstruction and prediction methods based on artificial intelligence have been introduced. These tools will allow dealing with more complex and nonlinear data in order to model optical aberrations more accurately than in the case of traditional methods. Due to this, enhancements in terms of precision and efficiency, and also higher ability to adapt to real-time changes of the atmospheric turbulence, are achieved. Furthermore, neural networks depend less on explicit physical models, and as will be shown in this work, they can reduce sensor noise and measurement errors. In total, the system becomes more robust against changes in atmospheric conditions.

The article builds upon two previous studies: firstly, the work of [15], which showcased enhanced prediction performance through the use of recurrent neural networks (RNN); secondly, [19], which employed a more complex model utilizing 2D-LSTM layers, resulting in increased efficiency and greater stability to changes in atmospheric turbulence parameters. The data obtained for the experiments will be based on atmospheric conditions similar to those adopted in both works, ensuring a faithful comparison with the results obtained in those studies. So far, the experiments conducted have not considered the possibility of removing sensor noise, which would allow for more direct predictions closely resembling a real wavefront. To address this, a noise parameter is introduced in the simulator, from which the data for training the predictive models will be generated. This parameter will be detailed further in Section 3 and will correspond to the noise component that can be eliminated. The primary objective of this work is to analyze the behavior of the prediction models in the presence of that noise.

This study expands upon the experiments conducted in, where the effect of wavefront sensor noise on future atmospheric turbulence prediction was briefly evaluated. In this new version, a more rigorous evaluation of the model's hyperparameters is performed to seek the best possible results. Additionally, two novel models are introduced: one based on the cascade prediction method, and an ensemble model that combines the strengths of the previously considered models to achieve optimal performance.

Section 2 delves into the basics and fundamentals of adaptive optics, along with the sources of noise that generate residual errors in wavefront measurements. Section 3 introduces the experimental background for the experiments conducted in this work, which are further described in this section. Finally, Section 4 offers a comprehensive analysis of the results obtained from these experiments.

2 Adaptive Optics

An AO system could be described as a set of tools developed to achieve high-quality imaging in large telescopes. It allows the manipulation of atmospheric disturbances with the goal of correcting the aberrations introduced in the wavefront captured by the measurement instrument. As explained above, the system is divided into two major subsystems. First, atmospheric turbulence in the environment is measured using Shack-Hartmann wavefront sensors. The slope acquisition, more commonly known as centroids is performed via the center of gravity approach. In this method, the centroid is computed as the weighted average of the position coordinates (x, y) [7], where the weighting function is the intensity of the spot as a function of position coordinates, I(x, y) = I(x, y):

$$(x_s, y_s) = \left(\frac{\sum I_{ij} X_{ij}}{\sum I_{ij}}, \frac{\sum I_{ij} Y_{ij}}{\sum I_{ij}} \right) \tag{1}$$

where i and j represent row and column indices, ranging from 1 to M, the size of an individual subaperture matrix (M × M).

In the second step, the system will compensate for the previously estimated distortions through a deformable mirror, which is able to adjust its surface in real-time based on the measurements from the sensor. [5].

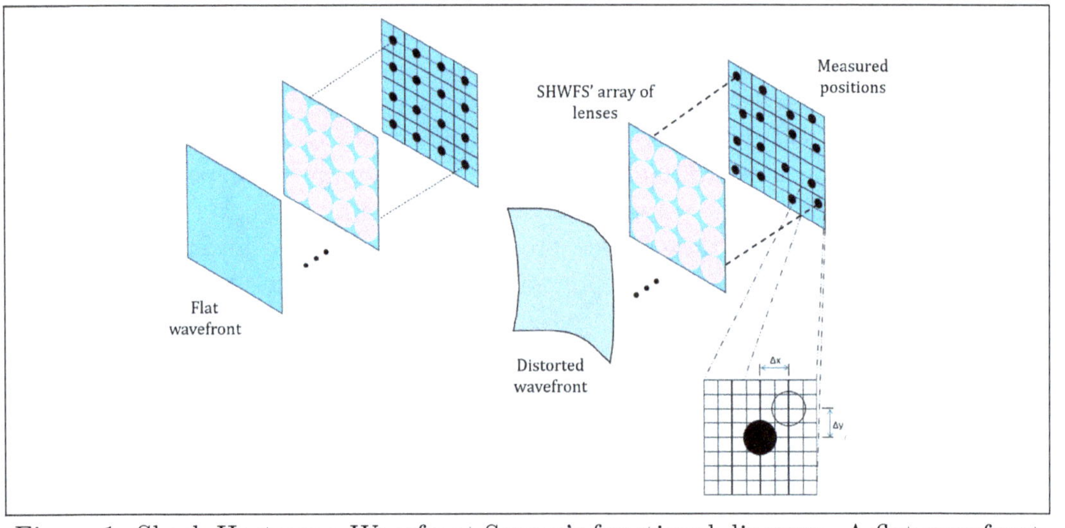

Figure 1: Shack-Hartmann Wavefront Sensor's functional diagram. A flat wavefront is shown on the left while a distorted wavefront is on the right. [1]

The distortions caused by atmospheric turbulence, being a stochastic phenomenon, can be characterized by the Kolmogorov model [27]. This model is based on Fried's parameter, which defines the coherent length scale over which wavefront distortions due to atmospheric turbulence remain roughly constant. It describes the maximum telescope aperture diameter for which resolution remains diffraction-limited despite turbulence. Fried's parameter is defined by the following Equation [6]:

$$r_0 = \left[0.423k^2 \sec(\zeta) \int C_n^2(z)dz\right]^{-3/5} \tag{2}$$

Here, ζ represents the zenith angle, which is the angular distance between the zenith and the astronomical object being observed. The wavenumber k, which depends on the wavelength λ of the observed light, is given by the following expression:

$$k = \left(\frac{2\pi}{\lambda}\right) \tag{3}$$

The parameter $C_n^2(z)$ represents the refractive index structure as a function of altitude z, providing a detailed characterization of atmospheric turbulence at various heights [10]. Local atmospheric conditions can be modeled by adjusting the parameters A and W in the following Equation [25]:

$$C_n^2(z)) = 5.94\times10^{-23}z^{10}\left(\frac{W}{27}\right)^2 exp(-z)+2.7\times10^{-16}exp\left(\frac{-2z}{3}\right)+Aexp(-10z) \quad (4)$$

The impact of atmospheric turbulence on the results obtained by telescopes varies depending on r_0 magnitude. When the value of r_0 is low, images captured may be badly distorted while with higher values, the distortion effects are largely reduced [4]. Improvement in image capture is yet a subject of continuous studies since, despite the advantages of space telescopes, ground-based telescopes are still indispensable for many scientific fields [8].

2.1 Simulator Soapy

For conducting experiments, a simulator developed by the University of Durham [21] is employed.

The simulator evaluates the light wavefront passing through the atmosphere in each iteration using the previously mentioned SH-WFS sensor. Subsequently, this data is processed by the DM's reconstruction modules, generating slopes that serve as inputs for prediction models.

After acquiring the slopes through Soapy simulator, it is necessary to obtain the primary Zernike aberrations. An essential step for later determining the residual wavefront error values. The Zernike polynomial series are preceded by Zernike coefficients, those are radial A_{nm} and azimuthal B_{nm} [25]. The Zernike polynomials are defined as:

$$Z_n^m(\rho,\theta) = \begin{cases} N_n^m R_n^{|m|}(\rho)\cos \mathrm{m}\theta & \text{for m} \geq 0, \\ -N_n^m R_n^{|m|}(\rho)\sin \mathrm{m}\theta & \text{for m} < 0, \end{cases} \quad (5)$$

where N_m^n is a normalization constant 6, being n and m indices that refer to each of the Zernike polynomials, while the first one corresponds to the radial order, the second one is the azimuthal frequency. On the other hand, component $R_{|m|}^n(\rho)$ relates to radial polynomials 7.

$$N_n^m = \sqrt{\frac{2n+2}{1+\delta_{0m}}} \quad (6)$$

$$R_n^{|m|}(\rho) = \sum_{s=0}^{(n-|m|)/2} \frac{(-1)^s (n-s)!}{s! \left[\frac{n+|m|}{2} - s\right]! \left[\frac{n-|m|}{2} - s\right]!} \rho^{n-2s} \quad (7)$$

To perform the transformation, a Zernike reconstruction matrix tailored for the CANARY configuration is employed. This matrix is applicable to a 7x7 Shack-Hartmann system operating on a circular aperture, the diameter of which has a ratio of 1.2/4.2. It provides Zernike coefficients for modes extending from tip to z=37, encompassing a total of 36 Zernike modes.

The simulation parameters are summarized in Table **??**:

Module	Parameter	Value
Atmosphere	No. phase screens	1
	Gain	1
	Wind speeds	from 10 to 15 m/s
	Wind direction	0–360 deg
	Screen height	from 1 to 11 km
	r_0@500 nm	0.16 m
	L_0	25m
Telescope	Diameter	4.2 m
	Central obscuration	1.2 m
SH-WFS	GS magnitude	10
	No. sub-apertures	7×7
	Readout noise	1 e^- RMS
	Photon noise	True
	Wavelength	600 nm
	Thresholding value	0.1
Piezo DM	No. actuators	8×8

Table 1: Main set of parameters for the Soapy SCAO simulation.

Through the Soapy simulator, two distinct datasets are generated for our experiments. For both datasets, the input slopes are subject to simulated sensor noise, mirroring real-world measurement conditions. However, a key distinction lies in their corresponding outputs. The "clean" dataset consists of output slopes where the simulated sensor noise has been intentionally removed during the data generation process, providing an ideal, noise-free representation of the atmospheric turbulence. Conversely, the "noisy" dataset maintains the simulated sensor noise in its outputs, reflecting the actual measurements that would be obtained from a sensor in a practical adaptive optics system.

2.2 Noise sources in AO

In the process of capturing aberrations through the wavefront sensor, various sources of noise are encountered [17]. So, while it is possible to train a model to predict measurements without noise, it will not completely learn exact behavioral patterns. In other words, it can reduce the error to some extent, but it can never entirely eliminate it in such predictions. That said, different types of errors can be found, [2], which are listed below:

- **Aliasing error:** It takes place when the spatial frequency of atmospheric turbulence exceeds the capability of the wavefront sensor. Following the corresponding approach of [22], it can be defined as:

$$\sigma^2_{aliasing} = 0.08 \left(\frac{d_{DM}}{r_0}\right)^{\frac{5}{3}} \tag{8}$$

  where d_{DM} the spacing between individual actuators.

- **Fitting error:** In this case, the source of error is caused by the deviation between the ideal wavefront correction and the actual correction carried out by deformable mirrors. Fitting error can be found as follows:

$$\sigma^2_{fitting} = a_F \left(\frac{d_{DM}}{r_0}\right)^{\frac{5}{3}} \tag{9}$$

  where a_F is a coefficient that depends on the class of actuator used.

- **Temporal error:** This source of error exists due to inaccuracies in the correction of optical systems that come up from delays between the wavefront sensing and the following implementation of corrections. Temporal error is defined as:

$$\sigma^2_{temporal} = \left(\frac{\tau_d}{\tau_0}\right)^{\frac{5}{3}} \tag{10}$$

  where τ_d is the respective time lag.

- **Noise error:** Two types of noise sources in adaptive optics systems need to be defined when simulating our data and conducting subsequent experiments:

- **Photon Noise:** The discrete nature of photons leads to a variable number of light particles reaching the detector, introducing a noise component into the images.
- **Readout Noise or RON:** Due to the inherent characteristics of CCD detectors, noise can be introduced in the reading of pixel values.

In the context of a Shack-Hartmann Wavefront Sensor (SH-WFS), the focal plane image, denoted as I_{WFS}, is subject to random contamination from both Poissonian photon noise and RON. In Equation 11, $\mathcal{P}$ represents a Poisson process, and n_e signifies the Root Mean Square of RON in terms of the quantity of electrons per CCD pixel [14].

$$I_{WFS} = \mathcal{P}(I_{WFS}) + n_e \tag{11}$$

On a more general level, the error of an adaptive optics system can be found as follows:

$$\sigma^2_{Total} = \sigma^2_{Aliasing} + \sigma^2_{Fitting} + \sigma^2_{Temporal} + \sigma^2_{Noise} \tag{12}$$

2.3 Neural Networks

Currently, neural networks have made their mark in virtually every scientific field, and their capabilities are undeniable when it comes to tasks such as pattern recognition [11], classification [3], and natural language processing [24], among others. These networks consist of artificial neurons that mimic the behaviour of those found in the human cerebral cortex, arranged in stacked layers.

A conventional neural network consists of an input layer that receives the initial data, one or more hidden layers responsible for processing information based on associated parameters (weights and biases), which are adjusted as iterations progress to learn specific task-related patterns. It also includes an output layer that generates the model's predictions. Supervised learning is employed, involving the use of a labeled dataset for training, validation, and testing of the neural network. Through this process, the model iteratively adjusts the mentioned parameters, progressively minimizing the difference between the output predictions and the actual labels corresponding to the labelled outputs [23]. In 2012, in [18], the feasibility of utilizing neural networks within the field of adaptive optics was demonstrated, achieving an effective wavefront reconstruction.

Although significant advances have been made in the application of neural networks to AO over the last two decades, it is in recent years that superior results have been achieved by incorporating more sophisticated deep learning architectures [16]. These structures can more efficiently capture the spatio-temporal relationships of atmospheric turbulence, as is the case with 2D-LSTM networks, which integrate convolutional operations to process spatial features while the recurrent structure models their temporal evolution [19]. More recently, other architectures such as Transformers have demonstrated remarkable results by employing attention mechanisms to evaluate the relevance of long-term past information, which allows for greater model efficiency at a reduced computational cost [28]. However, its application to real-world data has not yet achieved results as favourable as other options [26]. Another example, considering an encoder-decoder architecture, is the U-Net [9]. This model has demonstrated high efficacy in wavefront reconstruction from sensor data, owing to its ability to capture spatial features at multiple scales. Furthermore, the growing relevance of generative networks (GANNs) must be highlighted, both for wavefront reconstruction and future prediction [13], although it is necessary to properly analyse the higher computational cost they require [20]. Finally, it is worth highlighting Physics-Informed Neural Networks (PINNs), which integrate differential equations governing fluid dynamics into the model's loss function [12].

Despite the promising results shown by various proposed network architectures, neural networks based on 2D-LSTM layers were selected for this work. This choice is based on their strong capability in processing grid-like data, such as those from the considered wavefront sensor (Shack-Hartmann), which exhibit both spatial and temporal dependencies.

Furthermore, these networks allow for rapid training and good generalization, which facilitates the experimentation process given the time constraints involved in generating and processing large datasets of atmospheric turbulence under the specific configuration and conditions established in the simulation.

3 Experiments

In this work, a model previously studied in [19] is analysed, focusing on the feasibility of forecasting noise-free slopes.The relevant feature of this model is the existence of a certain kind of layer termed 2D-LSTM, which allows the analysis of temporal and spatial dependencies in a task containing sequential data.

The architecture used in the experiments is defined in Table 2 of the previous study [19]. It consists of a set of 2D-LSTM layers with BatchNormalization, followed by a section with non-spatial Conv2D layers. The aforementioned table provides the kernel size and the number of units established for each layer. Furthermore, an Adam optimizer is used, along with the "ReLU" activation function and the mean squared error ("mse") cost function. A learning rate reduction strategy is also employed using the "ReduceLROnPlateau" callback, with a factor of 1/5 and a patience of 5 epochs. The training is configured for a total of 40 epochs.

The number of available training data points is reduced to 40,000 from 100,000. This reduction is due to the necessity of separately generating noisy and clean data, as well as the increase in the number of atmospheric phases to 2 instead of the original 1 in the original work. These changes result in a significantly higher computational cost in the data simulation process.

To continue the examination of past experiments and facilitate a comparative analysis across various scenarios, the reference method for residual Wavefront Error outlined in [19] will be followed. More specifically, residual error obtained from specific centroids will be compared with the overall AO system error.

$$\text{Residual WFE } [\%]_{delayed} = \left[\frac{WFE_{zero-delay} - WFE_{delayed}}{WFE_{zero-delay}}\right] \tag{13}$$

$$\text{Residual WFE } [\%]_{predicted} = \left[\frac{WFE_{zero-delay} - WFE_{predicted}}{WFE_{zero-delay}}\right] \tag{14}$$

3.1 Feasibility of labelled clean data

There are two sets available: the set with noise and the set without noise. Every instance is produced from the same simulation and the same phase of atmospheric turbulence. Two training phases will be distinguished in this manner:

- **Noise-Clean Model (N-C):** A sequence size of 30 time-steps (s_0, s_1,..., s_{28}, s_{29}) will be maintained, and inputs will match the data from the noisy set. Regarding the output data, it will come from the corresponding next time-step, but imported from the clean set.

- **Noise-Noise Model (N-N):** The only difference compared to the previous case is that the outputs will be derived from the noisy set.

This approach allows for the observation of the model's behaviour when training with noisy data as the target, in contrast to using non-noisy data. Clean data without noise will never be considered in the input data since it wouldn't make sense in reality, as the values captured by the sensor will always have that associated noise error.

In this case, a direct multi-step prediction strategy is employed to predict the next 20 time steps from a sequence of the previous 30 time steps. A diagram of this process is shown in Figure 2. This approach involves training an independent and specialized model for each future time step. Since two scenarios are considered (one with noise and one without), this results in a total of 40 trained models. Furthermore, this methodology tests the accumulation of prediction errors and, although it is more computationally expensive, this strategy ensures that each forecast is generated independently from the original input sequence (ground truth), providing greater stability and precision.

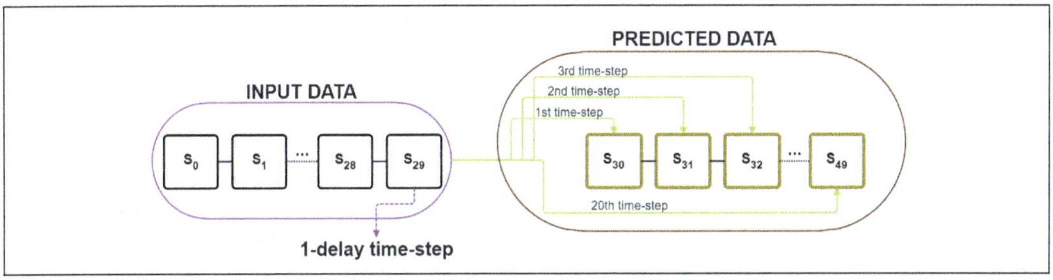

Figure 2: Representation of model training, input data and predictions.

3.2 Cascade prediction for clean time-steps.

To perform cascade prediction, the first frame of the future sequence is predicted. This prediction is then used as input for forecasting the next time-step, while the initial input data is removed to keep the input dimensions consistent. This process is repeated for each time-step until the entire sequence of predictions is obtained. The model used in the initial experiment predicts the first time-step s_{30} , and then reuses it for subsequent time-steps by incorporating both the input data and the predictions in a feedback loop.

The process can be visualized in Figure 3. In theory, the 1-delay reference would change in each iteration, as it represents the last available data prior to prediction (illustrated in the Figure). However, for the purpose of making predictions into the future from a specific point (with the first predicted frame being s_{30}), the 1-delay reference will be maintained consistently as the time-step s_{29}.

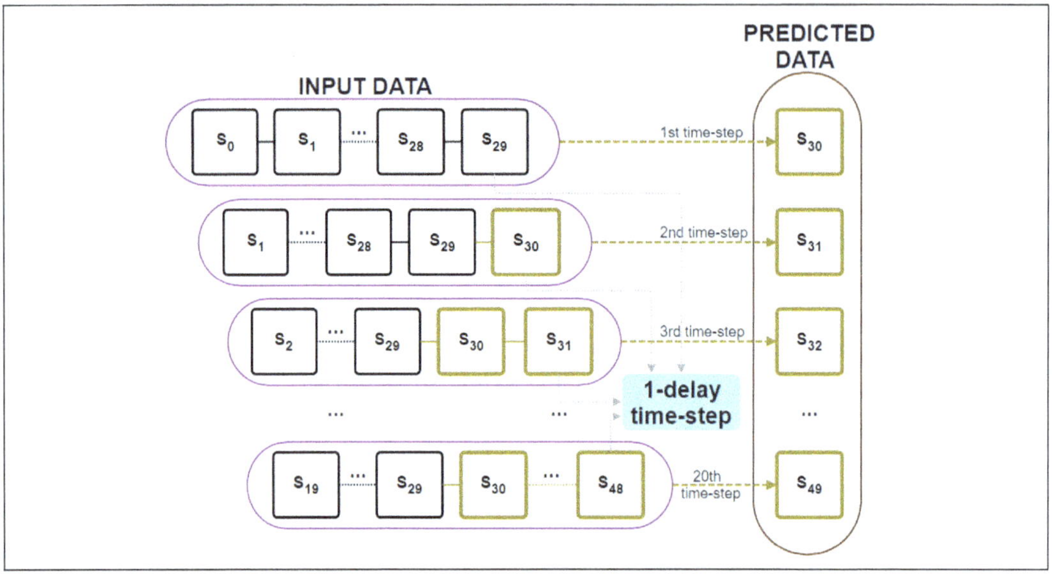

Figure 3: In-depth graphical overview of the cascade prediction technique. The fictional 1-delay reference of each prediction is shown, although the real one (s_{29}) will be considered for all of them.

3.3 Ensembling prediction without noise (N-C) and cascade method (N-Casc).

Given that the Noise-Clean experiment shows significant improvements for the first two time-steps and the cascade prediction experiment achieves a smaller yet noteworthy improvement for the next two time-steps, it was decided to build an ensemble of both models. This approach aims to achieve optimal error reduction across the first four frames of the predicted sequence. In this section, the ensembled model will be compared to the original Noise-Noise model.

In addition to the initial accuracy, assembling the two models would allow the flexibility and robustness of the cascade method to be exploited in the face of significantly unfavourable weather conditions, as well as taking advantage of the direct method to mitigate cumulative errors.

Additionally, an ensemble approach enhances model generalization, as it blends the strengths of both methods, minimizing the risk of overfitting to specific conditions.

4 Results

The performance evaluation of the models is conducted on multiple independent test datasets, each comprising 2000 sequences of 200 time-steps. For each test sequence, 200 consecutive predictions are generated. To ensure the model has reached a stable state and to avoid any initial transient effects, only the final 100 predictions from each sequence are considered for the analysis. This selection methodology guarantees that the evaluation focuses on the model's sustained performance and that the input values for prediction remain non-zero.

4.1 Feasibility of labelled clean data

The objective of this preliminary study is to evaluate the viability of using labelled data without noise as the neural network's output. Figure 4 shows the residual error results for the first four frames out of the twenty projected time-steps. For the first two time-steps in the N-C scenario, the system clearly performs better. This suggests that the neural network can learn strong features to reduce the inaccuracy caused by the noise in the system. Since the time-steps closest to the starting point of the future prediction are being considered, greater relevance is assigned to them compared to the rest. Unless the actual value matches by pure chance, forecast error increases with increasing distance from the reference point due to greater uncertainty. This might occur at a point long enough in the future to compromise prediction accuracy, but only considering an individual prediction.

Table 2 contains the residual errors in percentage for the three cases considered, as well as the error reductions that would be observed in each of the prediction cases (N-C & N-N) in relation to the reference (1-delay).

T	N-C error	N-N error	1-delay error	Reduction N-C	Reduction N-N
1	11.17 ± 0.21	12.25 ± 0.32	16.27 ± 0.51	31.34 ± 0.88	24.71 ± 0.41
2	13.26 ± 0.32	14.25 ± 0.39	17.90 ± 0.62	25.95 ± 0.83	20.38 ± 0.59
3	16.41 ± 0.43	16.39 ± 0.50	20.07 ± 0.65	18.25 ± 0.54	18.36 ± 0.18
4	18.53 ± 0.52	18.58 ± 0.62	22.47 ± 0.76	17.52 ± 0.48	17.32 ± 0.05

Table 2: N-C and N-N average residual error and proper reduction (in %) comparing to 1-delay case. Values are shown as mean $\pm$ standard deviation. First 4 time-steps are shown. The "T" column corresponds to the time step being considered.

Looking further ahead, for the third and subsequent time-steps both models achieve similar error reductions, up to a distant point where the N-C model shows

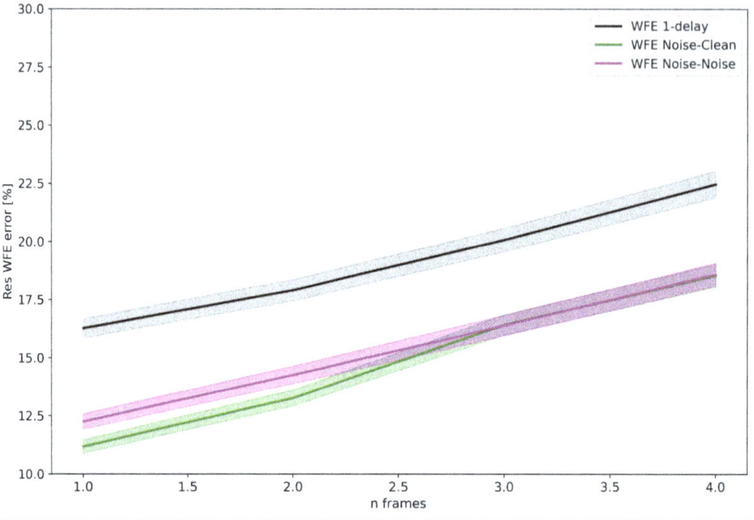

Figure 4: The residual error is presented for the three cases considered: 1-delay or reference (Grey), Noisy outputs (Blue), and Noise-free outputs (Green). The shaded regions indicate the standard deviation of the residual WFE.

some instability. However, long-term prediction is not of interest, as it is more efficient for the model to restart the process at short time intervals.

4.2 Cascade prediction for clean time-steps.

Based on the results shown in Figure 5, no significant reduction in errors is observed in comparison to the N-C case for the first couple of frames. Moreover, from the 4th time-step onward, the error rates between both models converge, showing no further discernible improvement. However, a notable enhancement is observed specifically in the 3rd and 4th time-steps, suggesting that this period holds potential for optimization. In the next experiment, the possibility of combining both prediction models will be explored in an effort to maximize performance during the first 4 time-steps.

Table 3 contains the residual errors in percentage for the three cases considered: 1-delay, Noise-Clean, and Cascade Prediction, as well as the error reductions that would be observed in each of the prediction cases (N-C & N-Casc) in relation to the reference (1-delay).

Taking a closer look at the potential reasons behind these results, it becomes clear that a similar outcome in the 1st frame is to be expected. This is because the first step in the cascade method mirrors the direct method, resulting in comparable performance at the initial stage. However, by the 2nd frame, where the direct method already performs quite well, achieving further improvement with the cascade method proves to be more challenging. As the analysis moves beyond the 2nd frame, the performance of the direct method starts to deteriorate, which allows the cascade method to step in and provide enhancement. Nonetheless, starting from the 4th frame, the errors accumulated in the cascade method start, as in previous models, to compound, leading to a noticeable decline in its effectiveness. This gradual increase of error over time begins to offset the advantages of the cascade method, particularly in longer predictions.

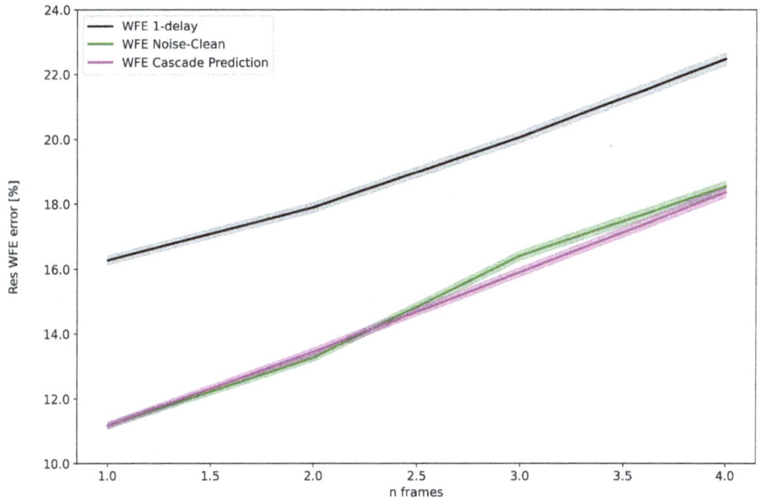

Figure 5: The residual error is presented for the three cases considered: 1-delay or reference (Grey), Cascade method (Blue), and Noise-free outputs (Green). The shaded regions indicate the standard deviation of the residual WFE.

T	N-Casc	N-C	1-delay	Reduction N-Casc	Reduction N-C
1	11.17 ± 0.07	11.17 ± 0.09	16.27 ± 0.12	31.34 ± 0.07	31.34 ± 0.05
2	13.45 ± 0.12	13.26 ± 0.13	17.90 ± 0.16	24.88 ± 0.02	25.95 ± 0.04
3	15.91 ± 0.15	16.41 ± 0.16	20.07 ± 0.21	20.75 ± 0.11	18.25 ± 0.07
4	18.36 ± 0.18	18.53 ± 0.19	22.47 ± 0.23	18.30 ± 0.04	17.52 ± 0.02

Table 3: N-Casc and N-C average residual error and proper reduction (in %) comparing to reference. Values are shown as mean ± std deviation. First 4 time-steps are shown. The "T" column corresponds to the time step being considered.

Time-step	Reduction N-Casc	Reduction N-C	Reduction N-N
1	**31.34**	**31.34**	24.71
2	24.88	**25.95**	20.38
3	**20.75**	18.25	18.35
4	**18.30**	17.52	17.32

Table 4: Error reduction comparison between three considered cases of training. Best average results are highlighted in bold.

4.3 Ensembling prediction without noise (N-C) and cascade method (N-Casc).

As explained before, the combination of both prediction models optimizes the results for the first four time-steps to a far more robust ensemble model leveraging the strengths of both approaches. It can be inferred from Table 5, which compares the residual error of the new model against the reference and the N-N model, that this constitutes a significant improvement over the latter.

T	ENS	N-N	1-delay	Reduction ENS	Reduction N-N
1	11.17 ± 0.21	12.25 ± 0.32	16.27 ± 0.51	31.34 ± 0.88	24.71 ± 0.41
2	13.26 ± 0.32	14.25 ± 0.39	17.90 ± 0.62	25.95 ± 0.83	20.38 ± 0.59
3	15.91 ± 0.15	16.41 ± 0.16	20.07 ± 0.21	20.75 ± 0.11	18.25 ± 0.07
4	18.36 ± 0.18	18.53 ± 0.19	22.47 ± 0.23	18.30 ± 0.04	17.52 ± 0.02

Table 5: Ensemble (ENS) and N-N average residual error and proper reduction (in %) comparing to reference case. Values are shown as mean ± std deviation. First 4 time-steps are shown. The "T" column corresponds to the time step being considered.

4.4 Results Overview

The results of this study confirm that the proposed models perform successfully in forecasting the noise-free slopes of an adaptive optics system.

First, two different models were considered: On the one hand, the Noise-Clean Model, or N-C, with noisy inputs and clean outputs, and on the other hand, the Noise-Noise Model, or N-N, which was based on noisy inputs and outputs. Then, the residual Wavefront Error, WFE, was computed with respect to the real value in order to quantify the accuracy of the prediction. From this, one might observe that residual errors from the N-C model were remarkably smaller for the first two time-steps, therefore, this model had learned and generalized well from the data.

Cascade prediction method has been considered as the second analysis, which did not bring significant improvements at the initial time-steps but showed potential improvements at third and fourth time-steps. However, the errors accumulated over time indicated some instability in long-term predictions.

The ensemble approach, which combined the strengths of both N-C and the cascade method, provided robust performance and lower residual errors compared to the N-N model. Overall, these results underscore the viability of using labeled clean data in training models for adaptive optics applications.

5 Conclusions and Future Lines

In this study, it has been explored the feasibility of forecasting noise-free slopes using a model based on 2D-LSTM layers within an adaptive optics system. With several experiments, it is demonstrated that the proposed Noise-Clean Model (N-C) significantly outperformed the considered model in prior studies, the Noise-Noise Model (N-N), in terms of residual Wavefront Error, particularly in the initial time-steps. The results indicate that the N-C model can successfully mitigate inaccuracies caused by noise, leading to improved predictive accuracy in atmospheric turbulence scenarios. In regards to the cascade prediction method, while exhibiting promising enhancements in specific time-steps, revealed challenges in long-term forecasting due to error accumulation. However, the ensemble approach that combined the strengths of both the N-C and cascade models effectively minimized errors, illustrating the potential of hybrid strategies in adaptive optics applications.

Future research will focus on several key areas to enhance the robustness and applicability of the developed models. First, it is expected to investigate several alternative neural network architectures, such as attention and transformer models, to assess their potential in capturing features in sequential data more effectively. Additionally, it would be useful to increase the complexity of the simulated atmosphere, which would improve model generalization across diverse operational scenarios. The mitigation of sensor noise and the strong performance shown in the different 2D-LSTM-based model approaches suggest a high potential for real-world application. This could lead to simpler, noise-free slope predictions, simplifying the wavefront correction process. Furthermore, it is intended to explore the real-time implementation of the developed models in practical adaptive optics systems. With the integration of these models in an optical bench, it would be posible to evaluate their performance in live conditions and refine them based on feedback from real-world applications.

References

[1] Andújar, J.M., González-Gutiérrez, C., Sánchez-Rodríguez, M.L., Calvo-Rolle, J.L., de Cos Juez, F.J.: Multi-gpu development of a neural networks based reconstructor for adaptive optics. Complexity **2018**, 5348265 (2018), https://doi.org/10.1155/2018/5348265

[2] Clénet, Y., Gendron, E., Gratadour, D., Rousset, G., Vidal, F.: Anisoplanatism effect on the e-elt scao point spread function. a preserved coherent core across the field. A&A **583**, A102 (2015), https://doi.org/10.1051/0004-6361/201425469

[3] Duda, R.O., Hart, P.E.: Pattern Classification and Scene Analysis. John Wiley & Sons (1973)

[4] Dutton, J.A.: Dynamics of Atmospheric Motion: (formerly the Ceaseless Wind). Dover, New York (1995)

[5] Freeman, R.H., Pearson, J.E.: Deformable mirrors for all seasons and reasons. Applied optics **21**(4), 580–588 (feb 1982). https://doi.org/10.1364/AO.21.000580

[6] Fried, D.L.: Statistics of a geometric representation of wavefront distortion. J. Opt. Soc. Am. **55**(11), 1427–1435 (Nov 1965). https://doi.org/10.1364/JOSA.55.001427, https://opg.optica.org/abstract.cfm?URI=josa-55-11-1427

[7] Fusco, T., Nicolle, M., Rousset, G., Michau, V., Beuzit, J.L., Mouillet, D.: Optimisation of a shack-hartmann-based wavefront sensor for xao system. Proc SPIE **5490** (10 2004). https://doi.org/10.1117/12.549562

[8] Gómez, S.L.S., González-Gutiérrez, C., Alonso, E.D., Santos, J.D., Rodríguez, M.L.S., Morris, T., Osborn, J., Basden, A., Bonavera, L., González, J.G.N., de Cos Juez, F.J.: Experience with artificial neural networks applied in multi-object adaptive optics. Publications of the Astronomical Society of the Pacific **131**(1004), 108012 (sep 2019), https://dx.doi.org/10.1088/1538-3873/ab1ebb

[9] Hu, R., Zhang, L., Liu, B., Xu, K., Zhang, X.: High-precision wavefront reconstruction network based on u-net and residual structure. Optics Express **31**(4), 6156–6169 (2023). https://doi.org/10.1364/OE.482084

[10] Hufnagel, R.: Propagation through atmospheric turbulence. In: The Infrared Handbook, chap. 6. USGPO (1974)

[11] Kohonen, T.: Self-organized formation of topologically correct feature maps. Biological Cybernetics **43**(1), 59–69 (1982), https://doi.org/10.1007/BF00337288

[12] Kovács, B., Szabó, G., Erdélyi, M.: Physics-informed neural networks for diffraction-free beam propagation. Optics Express **31**(19), 30693–30704 (2023). https://doi.org/10.1364/OE.499691

[13] Li, Y., Guo, H., Li, X.: Phasegan: a deep-learning-based generative adversarial network for atmospheric turbulence phase screen generation. Journal of the Optical Society of America A **40**(5), 939–946 (2023). https://doi.org/10.1364/JOSAA.484931

[14] Liu, X.: Wavefront Prediction Using Artificial Neural Networks for Adaptive Optics. Durham theses, Durham University (2021), http://etheses.dur.ac.uk/14076/

[15] Liu, X., Morris, T., Saunter, C., de Cos Juez, F.J., González-Gutiérrez, C., Bardou,

L.: Wavefront prediction using artificial neural networks for open-loop adaptive optics. Monthly Notices of the Royal Astronomical Society **496**(1), 456–464 (2020), https://doi.org/10.1093/mnras/staa1558

[16] Massey, F., Danolas, E., Charalambous, A., Loupas, A., Hadjigeorghiou, A., Pitris, C.: Deep learning for adaptive optics: a review. Journal of Physics: Photonics **5**(2), 022002 (2023). https://doi.org/10.1088/2515-7647/acb7f7

[17] Matson, C.L., Roggemann, M.C.: Noise reduction in adaptive-optics imagery with the use of supportconstraints. Appl. Opt. **34**(5), 767–780 (Feb 1995). https://doi.org/10.1364/AO.34.000767, https://opg.optica.org/ao/abstract.cfm?URI=ao-34-5-767

[18] Osborn, J., Juez, F.J.D.C., Guzman, D., Butterley, T., Myers, R., Guesalaga, A., Laine, J.: Using artificial neural networks for open-loop tomography. Opt. Express **20**, 2420–2434 (2012). https://doi.org/10.1364/OE.20.002420, http://www.opticsexpress.org/abstract.cfm?URI=oe-20-3-2420

[19] Pérez, S., Buendía, A., González, C., Rodríguez, J., Iglesias, S., Fernández, J., De Cos, F.J.: Enhancing open-loop wavefront prediction in adaptive optics through 2d-lstm neural network implementation. Photonics **11**(3) (2024). https://doi.org/10.3390/photonics11030240

[20] Rao, C., Pu, Z., Jiang, W.: Adaptive optics based on machine learning: a review. Opto-Electronic Advances **5**(11), 210136 (2022). https://doi.org/10.29026/oea.2022.210136

[21] Reeves, A.: Soapy: an adaptive optics simulation written purely in Python for rapid concept development. In: Marchetti, E., Close, L.M., Véran, J.P. (eds.) Adaptive Optics Systems V. Society of Photo-Optical Instrumentation Engineers (SPIE) Conference Series, vol. 9909, p. 99097F (Jul 2016). https://doi.org/10.1117/12.2232438

[22] Rigaut, F.J., Veran, J.P., Lai, O.: Analytical model for Shack-Hartmann-based adaptive optics systems. In: Bonaccini, D., Tyson, R.K. (eds.) Adaptive Optical System Technologies. Society of Photo-Optical Instrumentation Engineers (SPIE) Conference Series, vol. 3353, pp. 1038–1048 (sep 1998). https://doi.org/10.1117/12.321649

[23] Rumelhart, D.E., Hinton, G.E., Williams, R.J.: Learning representations by back-propagating errors. Nature **323**, 533–536 (1986). https://doi.org/10.1038/323533a0

[24] Salton, G., McGill, M.J.: A statistical approach to mechanized encoding and searching of literary information. Journal of the ACM **30**(1), 108–131 (1983). https://doi.org/10.1145/322154.322164

[25] Tyson, R., Frazier, B.: Field Guide to Adaptive Optics. Field Guide Series, SPIE Press (2004), https://books.google.es/books?id=9fpMmHVw6e8C

[26] Wong, A.L., van Dam, M.A., Gavel, D.T.: A review of machine learning methods for wavefront control. Journal of Astronomical Telescopes, Instruments, and Systems **9**(03), 030901 (2023). https://doi.org/10.1117/1.JATIS.9.3.030901

[27] Zilberman, A., Golbraikh, E., Kopeika, N., Virtser, A., Kupershmidt, I., Shtemler, Y.: Lidar study of aerosol turbulence characteristics in the troposphere: Kolmogorov and non-kolmogorov turbulence. Atmospheric Research **88**, 66–77 (04 2008). https://doi.org/10.1016/j.atmosres.2007.10.003

[28] Zou, Y., Zhang, Y., Liu, Z., Dong, L., Zhang, D.: DeTurb: Atmospheric turbulence mitigation with deformable 3D convolutions and 3D swin transformers. arXiv preprint arXiv:2403.11182 (2024)

Received 6 January 2025

Hybrid Computer Vision system for Traceability Improvement in the Metal Manufacturing Sector

Paula Arcano-Bea, Agustín García-Fischer, José-Antonio López-Vázquez, Elena Arce, Francisco Zayas-Gato
University of A Coruña, Department of Industrial Engineering, CTC, CITIC
{paula.arcano,agustin.garciaf,jose.lopez,elena.arce,f.zayas.gato}@udc.es

Antonio Abelha
Escola de Engenharia da Universidade do Minho, Algoritmi/Lasi, Campus de Azurém, 4804 - 533 Guimarães, Portugal
abelha@di.uminho.pt

Pablo Fariñas
University of A Coruña, Department of Naval and Industrial Engineering
pablo.farinas@udc.es

Abstract

The integration of digital technologies in the shipbuilding industry has become essential to improve efficiency and ensure precision in complex manufacturing processes. This study explores the application of computer vision techniques for the identification and traceability of minor and simple subassemblies in shipbuilding. We developed a dual approach combining 3D point cloud analysis and 2D deep learning based instance segmentation to detect and classify components. A depth camera was used to acquire 3D point clouds and high-quality 2D images. For the 3D approach, we used surface and edge matching techniques, and for the 2D approach, we fine-tuned advanced instance segmentation models such as YOLO11 and MaskRCNN2. The results show that this combined approach significantly improves the reliability of subassembly identification, which is essential for improving traceability, reducing errors, and optimizing production workflows in the shipbuilding sector.

Keywords: Computer vision, Shipbuilding, Surface Matching, Deep Learning, YOLO11, Mask R-CNN.

1 Introduction

Digital technologies have become a fundamental part of the modern manufacturing industry, transforming traditional production methods by optimizing processes and significantly improving operational efficiency [1]. By embracing these advances, industries can quickly access critical information and optimize resource usage, which are critical to remaining competitive in an increasingly dynamic global marketplace [2]. The ability to adapt to these innovations has allowed industries to not only improve efficiency, but also to meet increasing demands for precision and reliability in production.

In the context of shipbuilding, the adoption of digital technologies is particularly critical. Shipyards operate in a highly competitive market where continuous improvement in the quality of end products is essential to remain relevant [3]. Shipbuilding is inherently complex, requiring extensive coordination of limited resources over extended timelines, adding further challenges for maintaining cost effectiveness and operational efficiency [4]. To address these challenges, many European shipyards, including those in Spain, have turned to modernization strategies that rely strongly on subcontractors. These subcontractors play an indispensable role in the production chain, providing essential components for the final assembly of ships [5].

Among the critical phases of shipbuilding, block assembly is one of the most resource-intensive processes. It accounts for 40-60% of the total labor hours and 30-50% of the total hull construction cost [6]. Central to the success of this process is accurate assembly sequence planning (ASP) and accurate sub-assembly identification, which together provide the foundation for efficient and effective shipbuilding workflows [7]. ASP ensures that each step of the assembly process is carefully organized, minimizing production delays and reducing the likelihood of errors. Accurate subassembly identification allows shipyards to maintain full control over the thousands of components involved in the shipbuilding process.

Subassemblies are among the smallest structural units in shipbuilding, consisting of stable structures formed by two or more components [8]. They are fabricated in earlier stages of the assembly process and contribute to the construction of larger structures, such as hull sub-blocks. Proper assignment and management of subassemblies is essential to facilitate operations, reduce assembly time, and minimize overall manufacturing costs [9]. However, the high variability and similarity of these parts creates significant challenges in their identification and traceability, leading to inefficiencies and potential errors in the production chain.

This study focuses on addressing these challenges through the application of computer vision technologies, using both 2D and 3D approaches to enable accurate subassembly identification. The first method is based on point cloud analysis, using surface and 3D edge matching techniques to recognize components based on their geometric contours. The second uses deep learning models for instance segmentation,such as YOLO11[10] and Mask R-CNN[11], providing precise pixel-level masks that distinguish individual objects within a category. This dual approach ensures complementarity between the 3D contour-based method and the pixel-level accuracy of the 2D models, enabling not only object classification and localization, but also detailed shape and edge identification. By combining these techniques, the study aims to provide a reliable solution for the identification and traceability of minor and simple subassemblies.

This document is organized as follows: the next section introduces the case study, providing the context and challenges addressed in this research. This is followed by a detailed description of the proposed approach and methodology. The following sections present the experimental setup, outline the results obtained, and analyze their implications. Finally, the paper concludes with a discussion of the contributions of the study and potential directions for future research.

2 Case of Study

For this study, real-world data was provided by a company specializing in metal manufacturing and acting as a subcontractor to a shipyard. The data provided consisted of the blueprints of some small components known as minor and simple sub-assemblies. These components are manufactured by the company and shipped to the shipyard for final integration into the ship's construction.

The assembly of these components, called sub-assemblies, plays a significant role in forming larger units, called sub-blocks and blocks, which are fundamental to the shipbuilding process.

These minor and simple sub-assemblies represent some of smallest structural elements in the entire assembly and typically consist of a base plate welded to one or more smaller stiffeners[8].

Given the inherent complexity of ship assembly, it is essential for shipyards to be able to accurately track all the individual components that constitute each block of the ship. This emphasizes the necessity of robust traceability systems for auxiliary suppliers. Companies involved in this process must carefully document essential details about each component, including its material composition, as a variety of materials are used in production and their proper documentation ensures quality. Additionally, information regarding the person performing the welding, along with the specific methods used, is critical to quality assurance. Each piece must also be clearly marked with its final destination within the ship, specifying the intended block, sub-block and group to which it will belong.

Due to the large number of models and variations of these components, it is common to encounter parts that are very similar to each other. Therefore, unique identification of each part is essential to avoid confusion. Proper traceability not only ensures that the strict standards of shipbuilding are met, but also increases efficiency in managing large assemblies where precision and reliability are essential.

Even though these minor and simple sub-assemblies are among the smallest components of a ship, they are of considerable size. Even the smallest base plates are nearly a meter long and weigh over 11 kg. These characteristics make the handling of these components a challenge, which led us to replicate several of them using 3D printing.

To do this, we created computer-aided design (CAD) models of the components and scaled them down to create more manageable pieces. This approach allowed us to conduct multiple experiments in a controlled and easily adjustable environment while maintaining the correct shape of the actual components.

In Figure 1a we have an example of one of the company's blueprints of a base plate, and in Figure 1b there is the created CAD model of that piece.

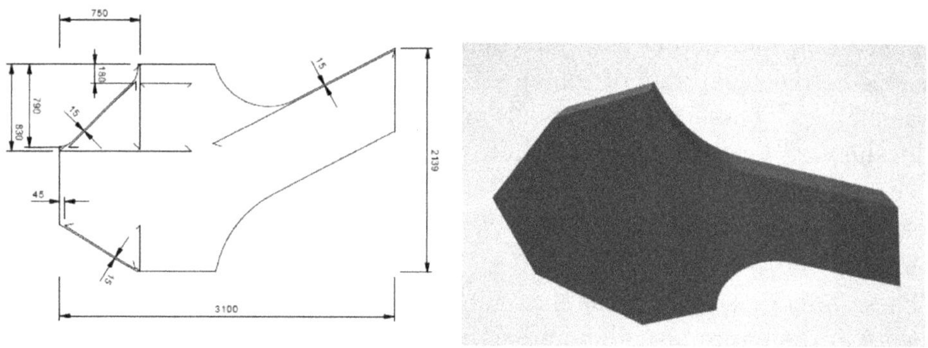

(a) Blueprint of a base plate. (b) CAD model of a base plate.

Figure 1: A blueprint of a base plate and the CAD model created with it.

3 Approach

To begin, we used the blueprints provided by a company acting as a subcontractor for a shipyard to create computer-aided design (CAD) models representing two minor subassemblies and three simple subassemblies. After completing the CAD models, we scaled them down to make them more manageable, with the largest component measuring 17 cm in length. These scaled-down models were then fabricated using a Fused Deposition Modeling (FDM) 3D printer, which creates objects by extruding a thermoplastic filament that melts and is deposited layer by layer.

For the material, we selected polylactic acid (PLA), a thermoplastic known for its ease of use, biodegradability, and sustainability. PLA is derived from renewable plant sources, making it an environmentally friendly option commonly used in 3D printing.

Specifically, we printed three simple subassemblies and two minor subassemblies to represent a range of typical components used in shipbuilding. These subassemblies were chosen for their simplicity and relevance to the manufacturing process, allowing us to focus on the identification and segmentation tasks.

To identify these components, we employed two different approaches. The first approach uses point cloud data, where we apply surface and 3D edge matching techniques to recognize components based on their geometric contours.

The second approach involves using 2D images processed through deep learning models. These models can be applied for various tasks, including object detection, semantic segmentation, and instance segmentation. Object detection locates objects within an image by enclosing them in bounding boxes, providing general information on their position and size. Semantic segmentation classifies each pixel in an image, assigning it to a particular category, but does not distinguish between individual objects of the same class. Instance segmentation overcomes this limitation by not only classifying pixels but also distinguishing between individual objects, generating precise pixel-level masks that accurately outline the contours of each instance.

For our 2D approach, we chose instance segmentation models, as they complement the 3D point cloud technique. This choice allows us to obtain not only the localization and classification of components but also their precise shape and edges, ensuring a high level of accuracy and consistency between the two methods.

To capture the necessary point clouds and images we used the Intel RealSense D455 depth camera. This camera is designed for precise 3D scanning with a range of 0.6 to 6 meters and a depth accuracy of less than 2% at 4 meters. Its stereoscopic imaging, using two synchronized infrared cameras, mimics human binocular vision to calculate distances based on pixel disparities. Additionally, structured light technology improves depth measurement by projecting infrared patterns, enabling accurate distance measurements even in varied lighting conditions. This data is processed into a depth map for precise 3D scene reconstruction.
This whole process is illustrated in Figure 2.

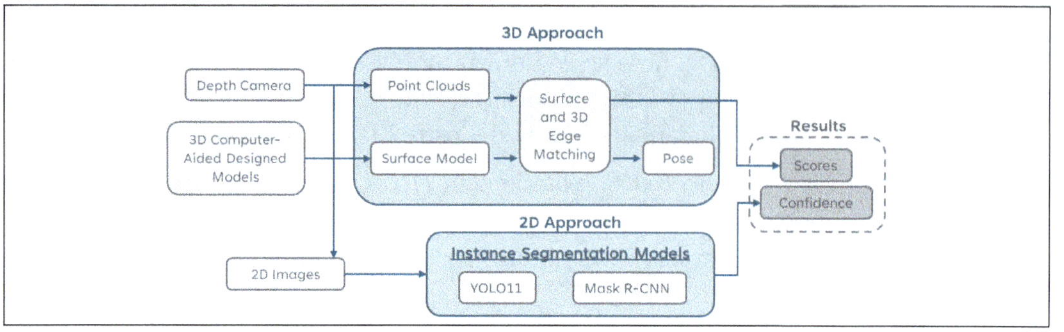

Figure 2: Visual representation of the approach.

4 Methods and materials

In this section, we are going to explain all the methods that were used in this study.

4.1 Dataset Creation

To create the dataset we capture more than eighty RGB images and point clouds of different scenarios using the Intel RealSense D455. For each scenario, we changed how many pieces it contained, as well as their orientation and position. Captures were made using the RealSense Viewer software, which allow us to visualize depth streams, RGB data, and infrared feeds and adjusts camera settings including depth and RGB resolution, frame rates, exposure, gain, and laser power. The software makes it easy for users to easily capture and save RGB and depth map images in PNG format and 3D images in PLY format.

The next step in creating our dataset is to label the 2D images. This labeling process involves manually annotating each image to create pixel-level masks for the components of interest. Using specialized annotation tools, we carefully traced the contours of each component, assigning a unique label to differentiate one instance from another, even when multiple objects belong to the same class.

4.2 Matching techniques

In this section we will explain all the techniques used to recognize the pieces.

4.2.1 Surface and 3d egde matching

Surface matching is performed in combination with three-dimensional edge matching. To start this process, it is necessary to create a surface model. A surface model is a triangulated point cloud that is an approximation of the true surface of an object[12]. Using a 3D object model as a reference, we create a surface model that can be trained to support multiple matching methods and refinement parameters, including 3D edge matching, view-based matching, self-similar pose training, and others. With these multiple training modes, we can achieve more accurate and reliable matching results. The matching process starts with the created surface models and the scene point cloud, where the system first identifies the approximate poses of the surface models within the scene. This is followed by a sparse pose refinement, where the initially identified poses are fine-tuned to improve the accuracy of the pose and the score value. The next step is the dense pose refinement which minimizes the distance between the scene points and the planes of the closest model points[13]. As a result of this matching process, we have the final pose and score for each element in the scene, ensuring precise alignment and accurate recognition. For this matching process we can generate up to five different types of scores, the first one is the surface score that measures the accuracy of the match between the surface model and the object's surface in the scene. The second one is the 3D edge score evaluates the quality of the match between the 3D edges of the surface model and the 3D edges of the scene object. The 2D edge score, that evaluates the match between the 2D edges of the surface model and the 2D edges of the object. The view-based score, which measures the quality of the match from different viewpoints. And finally, the global score, that which is the average of the scores, providing an overall assessment of the match quality. These scores range from 0.00 to 1.00, with 1.00 indicating a perfect match with 100% accuracy. The resulting pose of the matching process includes the X, Y, and Z coordinates, as well as the roll, pitch, and yaw angles.

The surface and 3D edge matching process was implemented using the HALCON software library developed by MVTec [14]. HALCON provides a framework for model-based 3D object recognition through surface-based and edge-based matching techniques. Its internal architecture includes efficient pose estimation pipelines and multi-stage refinement strategies (sparse and dense), which are optimized for industrial applications in real-time environments.

Although HALCON operates as a compiled library with optimized routines, inference times depend on the size of the point clouds, model complexity, and available hardware. Under standard conditions with mid-range industrial GPUs or CPUs, typical inference times range from 150 ms to 300 ms per object [15].

4.2.2 YOLO11

YOLO11, an advanced version of the You Only Look Once (YOLO) family of deep learning models. It enhances the established strengths of previous YOLO versions

[16], incorporating several advanced features that improve performance across a wide range of computer vision tasks. It introduces an optimized backbone and neck architecture that significantly improves feature extraction capabilities, allowing for more accurate object detection and more efficient handling of complex tasks. In addition, YOLO11 achieves higher accuracy with fewer parameters, resulting in a higher Mean Average Precision (mAP) on the COCO dataset while using 22% fewer parameters compared to YOLOv8m. This computational efficiency enables the model to maintain high performance in diverse environments, whether deployed on edge devices, cloud platforms, or systems using NVIDIA GPUs. YOLO11's adaptability ensures that it can handle a wide range of tasks, from basic object detection and instance segmentation to more specialized applications such as image classification, pose estimation, and oriented object detection (OOD), making it a versatile tool for various computer vision challenges[10]. There are several YOLO11 models such as YOLO11-n,YOLO11-s, YOLO11-m, YOLO11-l, and YOLO11-x, ranging in size from small to to large to suit different application scenarios[17]. YOLO11 integrates instance segmentation by predicting not only the class and bounding box of each object, but also the exact pixel-level mask for each instance. This capability is crucial for applications that require high precision in object identification, such as manufacturing and robotics, where distinguishing between different components is crucial. YOLO11 maintains the core architecture of the YOLO series, which consists of three primary components. The first is the Backbone, which acts as the primary feature extractor, dedicated to capturing essential features from the input image. The second is the Neck, which connects the Backbone and the Head and acts as a feature fusion module, combining multi-scale information from different layers of the network. And finally, the Head, which transforms the processed feature maps into precise object detection outputs, including object classifications and bounding box coordinates[18].

In this study, we used the YOLO11-m variant, which offers a balance between performance and computational cost. This model has approximately 25 million parameters, making it suitable for real-time applications while maintaining high detection accuracy [19] [20]. Regarding inference speed, YOLO11-m can achieve real-time performance with inference times of approximately 7–10 ms per image on an NVIDIA RTX 3080 GPU under standard conditions [21]. These times may vary depending on hardware and deployment environment.

4.2.3 Mask R-CNN

Mask Region-Based Convolutional Neural Network (Mask R-CNN) is a deep learning model designed for both object detection and instance segmentation tasks, known for its high precision and efficiency in generating accurate segmentation masks for each detected object. Mask R-CNN builds on the Faster R-CNN architecture by adding a parallel branch specifically for predicting segmentation masks alongside the existing bounding box prediction branch. This additional mask prediction capability enhances the model's ability to identify and delineate individual objects within an image[11]. The architecture of Mask R-CNN consists of three core components: the backbone network, the region proposal network (RPN), and two parallel branches.

The backbone, typically a convolutional neural network (CNN), extracts essential features from the input image that are shared by both the bounding box and mask prediction branches. The RPN generates region proposals that are likely to contain objects based on the feature maps from the backbone. The bounding box branch then classifies each region and predicts the bounding box coordinates, while the mask branch generates a binary mask for each object instance within the bounding box[22].

To improve the accuracy of segmentation, Mask R-CNN introduces RoI Align, a technique that improves the alignment of regions of interest (RoIs) and ensures finer segmentation accuracy, even in cases involving multi-scale targets [23]. This makes it well suited for challenging scenarios such as varying object scales and orientations, which are common in environments such as shipbuilding where parts may appear from different angles or have similar shapes.

For Mask R-CNN, we employed a configuration with ResNet-50 as the backbone, which includes approximately 44 million parameters [24]. This model is widely used due to its solid performance in instance segmentation tasks. In terms of inference time, standard implementations of Mask R-CNN with ResNet-50 typically achieve 80–120 ms per image on GPUs like the NVIDIA RTX 3080 [25].

5 Experiments and results

In this section, we will describe the experiments that were performed and the results of those experiments.

5.1 Experiment Setup

For our first approach, we used MVTec's HALCON software, a versatile machine vision tool known for its robust performance in image processing, machine vision, and 3D vision applications. HALCON's capabilities allowed us to efficiently handle complex point cloud and object recognition tasks. As described in the Methods and Materials section, the process began with creating surface models using HALCON's "create surface model" function, which enables training on multiple matching techniques such as 3D edge matching, surface matching, and view-based matching. This feature was very important, given the different orientations that objects can assume within the scene. The matching process involved several parameters to optimize performance. We set the minimum score threshold to 0.8 (80% confidence) to filter out low-quality matches, accounting for noise and lighting variations. The 3D edge extraction threshold was set to 0.01, ensuring that only significant edges were identified. To refine this, we used HALCON's graphical inspection tool, balancing detail retention with computational efficiency. For the relative sampling distance, we tested various values as recommended by HALCON's documentation [26], ultimately selecting 0.04, which offered a good balance between processing speed and accuracy. For the distance threshold relative to the surface model's diameter, we experimented with values from HALCON's documentation [26], starting from 0.1 down to 0.03. Our tests showed that lower values, particularly 0.03, improved results by focusing on points closer to the model (points within 3% of the object's diameter), ensuring better match quality.

For the 2D approach, we used consistent configurations for both deep learning models (YOLO11 and Mask R-CNN) to ensure a fair comparison of their performance. The configurations were chosen to optimize the training and testing conditions for the given task of sub-assembly identification and instance segmentation.The models were trained for 100 epochs to allow the model to adequately learn from the data without overfitting. We set the learning rate to 0.001 based on empirical tests and literature [22] recommendations for similar tasks, aiming to balance model accuracy with convergence speed. A batch size of 8 was chosen to train both models for memory efficiency and faster updating of the model weights, and also to ensure that the training process could be completed effectively given the computational resources available. The dataset was split into two parts: 80% for training and 20% for validation. This split is a standard practice, to ensure that there is enough data for training the model while retaining a separate portion for evaluating how well the model generalizes to new, unseen data.

All experiments were performed on a workstation running Windows 11, and an NVIDIA RTX 4078Ti GPU. The 3D matching tasks were implemented using MVTec HALCON, a commercial software widely used for industrial vision applications. For the 2D experiments, we used Python 3.11. The YOLO11 model was implemented using the official Ultralytics codebase [10], and the Mask R-CNN model was built using the open-source implementation [27]

5.2 Results and analysis

For our 3D approach, we conducted numerous test that involved creating surface models from all available CAD models and applying the matching process to different scenarios where the elements present in the scene were rotated, moved, and changed. It is important to emphasize that in order to easily identify our pieces and to avoid using the same references as those in the company's original blueprints for privacy reasons, we assigned a letter to each piece, ranging from A to E. This method also helped make the results easier to interpret.In Figures 3 we have two results obtained when applying surface and 3D edge matching techniques to the captured scene point cloud. In that figures we have in green a representation of the point cloud looks when it is zoomed out, and in red is the matching result for one of the pieces. In the results obtained in this scenario for Figure 3a(Piece B) we can see that the total score was 0.96 and the surface score was 0.91 and for Figure 3b(Piece E) these scores were 0.95, 0.96 and 0.93 respectively. For all the experiments performed, we calculated the arithmetic mean of the scores, which resulted in the following: the average total score was 0.938, the surface score was 0.953, the 3D edge score was 0.861, and the view score was 1.00. As we can see, the view-based score consistently achieved a perfect score of 1.00, indicating that the camera alignment reliably matched the expected reference view.

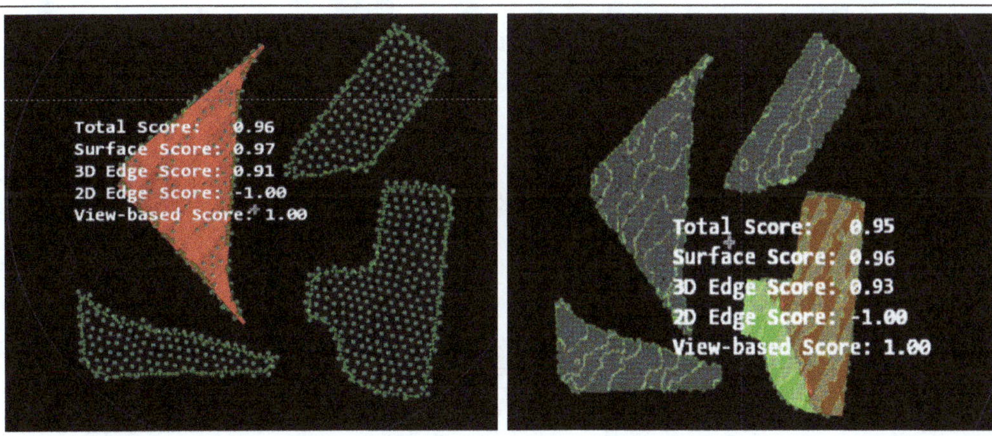

(a) Matching results for piece "B". (b) Matching results for piece "E".

Figure 3: Matching result of a scene.

For our 2D approach, we employed YOLO11 and Mask R-CNN to identify and classify the sub assemblies in the captured images. Both models gave very good results, as shown in the confusion matrices in Figures 4. The confusion matrices are a useful tool for evaluating the performance of classification models, showing the number or percentage of correct and incorrect predictions across all categories. They provide a detailed breakdown of how well the model distinguishes between different classes. In the x-axis we have the real class and in the y-axis the predicted class labels. The diagonal elements represent correct predictions, while the off-diagonal elements correspond to misclassifications. The higher the values along the diagonal, the better the model's accuracy in identifying and classifying the objects. For both of or model we created a normalized confusion matrix. For YOLO11 (Figure 4a), we achieved accuracies of 0.98 for piece A, 0.97 for piece B, 0.98 for piece C, 0.97 for piece D, and 0.98 for piece E. The model also misclassified four background instances, with two detected as piece E and two as piece C. For Mask R-CNN (Figure 4b), we achieved the following accuracies: 0.97 for piece A, 0.97 for piece B, 0.95 for piece C, 0.96 for piece D, and 0.97 for piece E. In terms of background detection, out of 10 background instances, three were incorrectly classified as piece C, three as piece D, and four as piece E.

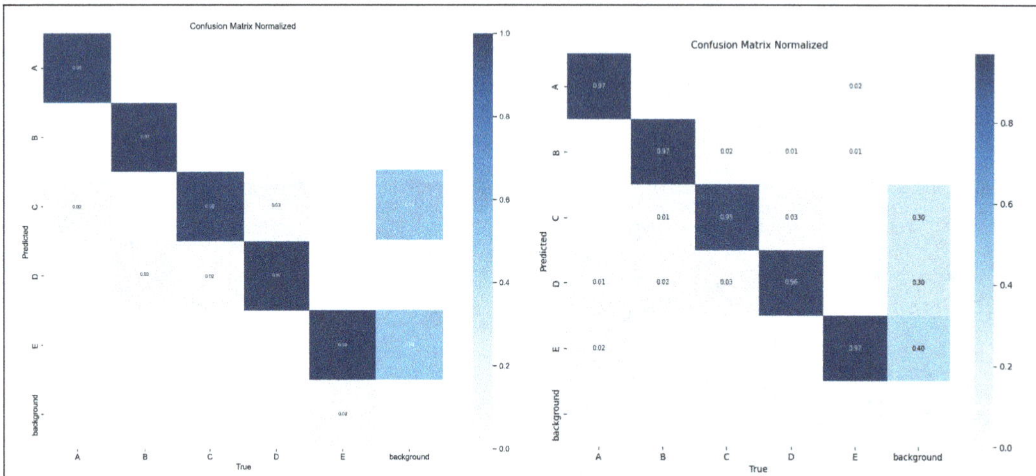

(a) Confusion matrix for YOLO11. (b) Confusion matrix for Mask R-CNN.

Figure 4: Confusion matrices for YOLO11 and for Mask R-CNN.

6 Conclusion and future works

The application of 3D and 2D approaches for object detection has become very useful in multiple fields, ranging from manufacturing to robotics and logistics. These methodologies are highly valued for their ability to deliver accurate and efficient results in complex scenarios, making them integral to advancing automated workflows. In the context of our study, we achieved satisfactory results for both the 3D and 2D approaches. For the 3D method, we observed strong performance with the surface score and overall total score being 0.953 and 0.938, respectively, indicating the effectiveness of this technique in identifying components under varying orientations and conditions. However, the 3D edge score was slightly lower at 0.861, suggesting a potential area for improvement in edge detection and matching. For the 2D approach, both deep learning models (YOLO11 and Mask R-CNN) had good results. YOLO11 consistently achieved high accuracies across all pieces, with scores between 0.97 and 0.98, and demonstrated efficient background identification with minimal misclassifications. Similarly, Mask R-CNN produced competitive results, with accuracy values ranging from 0.95 to 0.97 for all pieces, though it faced more challenges in distinguishing background elements, misclassifying 10 instances as subassembly components. Overall, both models had excellent performance, with YOLO11 achieving slightly higher accuracy values in most cases. Considering the strengths and limitations of both approaches, a hybrid model that combines the precision of the 2D deep learning models with the spatial accuracy of the 3D approach could offer significant benefits. Such a system could leverage the complementary strengths of each technique, ensuring both high accuracy in detection and robust spatial understanding. Such a system could combine the complementary strengths of each technique, using the precision and spatial awareness of the 3D approach alongside the detailed identification and classification capabilities of the 2D models. This integration would enhance accuracy in detecting components while also providing a clearer understanding of their spatial arrangement and orientation within the environment.

For future work, several path can be explored to improve the performance and applicability of this study. One potential improvement is the use of a constant artificial lighting condition to improve 3D edge matching, which addresses some of the challenges associated with varying illumination. Additionally, the implementation of a hybrid model could optimize the processing time for point cloud scenes by focusing only on the results provided by the deep learning model. This hybrid approach would not only reduce scene processing but also provide a double confirmation mechanism for the detected components, improving reliability. Another promising direction is to extend this methodology to real, full-scale components, to ensure that the models are effective in practical, industrial scenarios. Finally, expanding the training dataset to include a larger variety of piece types would further increase the system's versatility and applicability across multiple shipbuilding contexts.

Acknowledgments

This work has been supported by Xunta de Galicia through Axencia Galega de Innovación (GAIN) by grant IN853C 2022/01, Centro Mixto de Investigación UDC-NAVANTIA "O estaleiro do futuro", which is ongoing until the end of September 2025. The support was inherited from both the starting and consolidation stages of the same project throughout 2015-2018 and 2018-2021, respectively. This stage is also co-funded by ERDF funds from the EU in the framework of program FEDER Galicia 2021-2027.

Xunta de Galicia. Grants for the consolidation and structuring of competitive research units, GPC (ED431B 2023/49).

CITIC, as a center accredited for excellence within the Galician University System and a member of the CIGUS Network, receives subsidies from the Department of Education, Science, Universities, and Vocational Training of the Xunta de Galicia. Additionally, it is co-financed by the EU through the FEDER Galicia 2021-27 operational program (Ref. ED431G 2023/01)

References

[1] S. Zhao, X. Zhang, Did digitalization of manufacturing industry improved the carbon emission efficiency of exports: Evidence from china, Energy Strategy Reviews 57 (2025) 101614. doi:https://doi.org/10.1016/j.esr.2024.101614.

[2] D. Lu, E. C. M. Hui, J. Shen, J. Shi, Digital industry agglomeration and urban innovation: Evidence from china, Economic Analysis and Policy 84 (2024) 1998–2025. doi:https://doi.org/10.1016/j.eap.2024.10.047.

[3] S. Bertagna, L. Braidotti, V. Bucci, A. Marinò, Laser scanning application for the enhancement of quality assessment in shipbuilding industry, Procedia Computer Science 232 (2024) 1289–1298, 5th International Conference on Industry 4.0 and Smart Manufacturing (ISM 2023). doi:https://doi.org/10.1016/j.procs.2024.01.127.

[4] N. P. Basán, M. E. Cóccola, A. García del Valle, C. A. Méndez, An efficient milp-based decomposition strategy for solving large-scale scheduling problems in the shipbuilding industry, Optimization and engineering 20 (2019) 1085–1115.

[5] R. Mickeviciene, Global competition in shipbuilding: trends and challenges for europe, The Economic Geography of Globalization (2011) 201–222.

[6] S. Qu, Z. Jiang, N. Tao, An integrated method for block assembly sequence planning in shipbuilding, The International Journal of Advanced Manufacturing Technology 69 (2013) 1123–1135.

[7] C. Li, W. Hou, Assembly sequence planning based on hierarchical model, Wireless Communications and Mobile Computing 2022 (1) (2022) 9461794.

[8] B. Liu, R. Li, J. Wang, Y. Liu, S. Li, Subassembly partition of hull block based on two-dimensional pso algorithm, Journal of Marine Science and Engineering 11 (5) (2023) 1006.

[9] H. Cao, R. Mo, N. Wan, F. Shang, C. Li, D. Zhang, A subassembly identification method for truss structures manufacturing based on community detection, Assembly Automation 35 (3) (2015) 249–258.

[10] G. Jocher, J. Qiu, Ultralytics yolo11 (2024).
URL `https://github.com/ultralytics/ultralytics`

[11] K. He, G. Gkioxari, P. Dollár, R. Girshick, Mask r-cnn, in: 2017 IEEE International Conference on Computer Vision (ICCV), 2017, pp. 2980–2988. `doi:10.1109/ICCV.2017.322`.

[12] X. Li, J. Chen, H. Zhu, A new method for automated discontinuity trace mapping on rock mass 3d surface model, Computers & Geosciences 89 (2016) 118–131. `doi:https://doi.org/10.1016/j.cageo.2015.12.010`.

[13] A. Tellaeche, I. Maurtua, 6dof pose estimation of objects for robotic manipulation. a review of different options, in: Proceedings of the 2014 IEEE Emerging Technology and Factory Automation (ETFA), 2014, pp. 1–8. `doi:10.1109/ETFA.2014.7005077`.

[14] MVTec Software GmbH, Halcon – the comprehensive standard software for machine vision, available at: `https://www.mvtec.com/products/halcon` (2024).

[15] M. Schmidt, T. Becker, Benchmarking industrial 3d matching with halcon, Machine Vision Applications in Industry 9 (2023) 32–39.

[16] A. Sharma, V. Kumar, L. Longchamps, Comparative performance of yolov8, yolov9, yolov10, yolov11 and faster r-cnn models for detection of multiple weed species, Smart Agricultural Technology 9 (2024) 100648. `doi:https://doi.org/10.1016/j.atech.2024.100648`.

[17] D. Wang, J. Tan, H. Wang, L. Kong, C. Zhang, D. Pan, T. Li, J. Liu, Sds-yolo: An improved vibratory position detection algorithm based on yolov11, Measurement (2024) 116518`doi:https://doi.org/10.1016/j.measurement.2024.116518`.

[18] L. He, Y. Zhou, L. Liu, J. Ma, Research and application of yolov11-based object segmentation in intelligent recognition at construction sites, Buildings 14 (12) (2024). `doi:10.3390/buildings14123777`.

[19] Ultralytics, Yolov11: Next-generation real-time object detectionAccessed via official YOLO GitHub repository (2024).
URL `https://github.com/ultralytics/YOLOv11`

[20] L. Chen, J. Wang, A comparative study of yolov4, v5, v7 and v11 for real-time detection, Journal of Computer Vision Research 12 (2024) 45–60.

[21] A. Singh, P. Kumar, Yolov11 benchmarking on edge and cloud gpus, International Conference on Machine Vision (2024).

[22] R. Sapkota, D. Ahmed, M. Karkee, Comparing yolov8 and mask r-cnn for instance segmentation in complex orchard environments, Artificial Intelligence in Agriculture 13 (2024) 84–99.

[23] M. Yuan, H. Meng, J. Wu, S. Cai, Global recurrent mask r-cnn: Marine ship instance segmentation, Computers & Graphics 126 (2025) 104112.

[24] K. He, G. Gkioxari, P. Dollár, R. Girshick, Mask r-cnn, Proceedings of the IEEE International Conference on Computer Vision (2017) 2961–2969.

[25] F. A. Research, Detectron2 model zoo and baselines (2020).
URL https://github.com/facebookresearch/detectron2/blob/main

[26] M. Software, Halcon operator reference 21.05.0.0, https://www.mvtec.com/doc/halcon/2105/en/find_surface_model.html [Accessed: (22/11/2024)].

[27] W. Abdulla, Mask r-cnn for object detection and instance segmentation on keras and tensorflow, https://github.com/matterport/Mask_RCNN (2017).

 Received 10 January 2025

Smart Urban Pollution Management Using Deep Learning Techniques for NO_2 Prediction

Pedro Oliveira*, Afonso Bessa, João Pereira, Manuel Rodrigues
ALGORITMI/LASI Centre, University of Minho, Department of Informatics
{pedro.jose.oliveira,pg53597,pg52686,manuel Rodrigues}@algoritmi.uminho.pt

Francisco-Javier Pérez-Castelo, Andrés-José Piñón-Pazos, María-Carmen Meizoso-López
University of A Coruña, CTC, Department of Industrial Engineering
{francisco.javier.perez.castelo,andres.pinon,carmen.meizoso}@udc.es

Abstract

The rapid pace of urbanization and the expansion of economic activities have significantly contributed to rising levels of air pollution, posing serious threats to human health and the environment. Among the key pollutants, Nitrogen Dioxide (NO_2) stands out due to its strong association with respiratory and cardiovascular diseases and its role in exacerbating climate change. NO_2 is predominantly emitted through fossil fuel-powered transportation and industrial processes. At the economic level, air pollution entails high costs, such as lost productivity in key sectors and increased public spending on environmental remediation initiatives. In this context, Artificial Intelligence (AI) tools, such as Machine Learning (ML) and Deep Learning (DL), are now indispensable for analyzing environmental data and offering accurate predictions about pollution levels. Hence, this work focuses on the design, tuning, and evaluation of five DL models, namely Multi-Layer Perceptron (MLP), LSTNet, Temporal Convolutional Networks (TCN), Transformers, and Transformers combined with Long Short-Term Memory (LSTM), to forecast NO_2 concentrations in the city of Porto up to two days in advance. The results reveal that the MLP model demonstrated the highest performance, achieving a Root Mean Square Error (RMSE) of 7.65 $\mu g/m^3$, outperforming more complex architectures. These findings underscore the effectiveness of DL in pollutant forecasting, contributing to

*Corresponding Author

more informed decision-making and air pollution mitigation strategies.

Keywords: Deep Learning, Environmental Sustainability, Nitrogen Dioxide, Urban Air Quality, Time Series Problem

1 Introduction

Over the last few decades, the world has experienced significant growth in terms of population, doubling from 1974 (4 billion) to 2023 (8 billion), estimating a peak in the mid-2080s [1]. According to this marked increase, economic activities have also been increased, with more frequency of transport of goods and people by land, air and sea. Despite the significant evolution of many technologies involved in the transport sector, such as telecommunications or navigation systems, this sector still has a strong dependency on fossil fuels [2]. This dependency represents this sector and affects other activities, such as industry. It is an essential source of greenhouse gas emissions, causing climate change with potentially devastating and irreversible effects, such as temperature and precipitation changes or extreme climate events [3]. According to [4], more than 90% of transport technologies use fuel as an energy source, corresponding to 49% of the global share of oil consumption.

Apart from the climate emergency derived from the scenario described above, other air pollutants also have a substantial impact on human health, such as Carbon Monoxide (CO) and Nitrogen Dioxide (NO_2) [5]. With the focus on NO_2, it can be emitted through outdoor sources such as fossil fuel-powered vehicles and industrial or high-temperature combustion processes. It can also be generated in indoor processes, like smoking or gas stoves [6]. Furthermore, the generation of NO_2 is obtained from other Nitrogen oxides like NO, but it combines with Ozone to generate NO_2. The high level of exposure to this gas increases the possibility of having lung cancers and is directly correlated to other diseases in the liver, heart and kidneys [7, 8]. In [9], a strong correlation between some COVID-19 variants and NO_2 pollution is demonstrated. Although the industry has a lower impact since it is commonly placed far from urban centres, road transport is responsible for 25 % of all damaging emissions in Europe, contributing to 225.000 deaths by pollution-related diseases [8]. According to the study [10] developed by the World Health Organization (WHO), a global reduction in air pollution would increase life expectancy by 0.6 years. Besides the impact on human lives, NO_2 also affects the environment through ozone layer reduction, acid rain, biodiversity loss, and climate change [11].

Due to the importance of air pollution, countries worldwide are trying different ways to palliate the situation. As a result, the development of innovative and advanced energy technologies focuses on improving fuel usage efficiency and reducing

CO and NO_2 emissions [11]. Also, many cities implemented monitoring points at many points to have real-time information about air quality [12]. Different gasses can be measured like sulfur dioxide (SO_2), carbon dioxide (CO_2), NO_2, among others [10], using a wide range of technologies and sensors [13]. Real information about the current pollutant levels in different heat zones can contribute to identifying hazardous events. Besides the importance of air pollution monitoring, an accurate prediction of different pollutants' evolution can help anticipate dangerous situations, enhancing good decision-making [13]. In [14], authors propose the use of a time series approach to forecasting the NO_2 emitted by a Combined Cycle Power Plant using a Long Short Term Memory (LSTM) Neural Network combined with Multi-verse Optimization (MVO) algorithm. The prediction considers meteorological variables and lagged values of NO_2 to estimate future concentrations. The work presented in [15] proposes using Recurrent Neural Networks (RNNs) combined with meteorological parameters to determine future levels of six different pollutants in Skopje. Focusing on NO_2 forecasting, [16] proposes using Vector Autoregressive Moving Average (VARMA), Kriging, and seasonal Autoregressive Moving Average (ARMA) to determine concentration levels in London. Other works like [17] use clustering methods to determine NO_2 concentration probability distribution in different zones of Wroclaw.

Also related to pollution in urban areas, in [18], an analysis of different parameters affecting the NO_2 pollutants in Tehran indicates that parking areas present significant concentration levels. The authors deal with a regression problem, using Multi-Linear Regression (MLR) and Multi-Layer Perceptron (MLP) to determine the current NO_2 level from information about traffic activity, meteorological variables and time parameters. The study considers many variables that, in many cities or environments, are not feasible to register, such as traffic cameras. Furthermore, the possibility of forecasting future NO_2 concentration levels is not considered. A similar approach can be found in [19], where NO_2 and other pollutants concentration levels in big-sized cities by training an Artificial Neural Network with ambient temperature, humidity, wind speed and previous concentrations.

This work proposes forecasting NO_2 concentration using only NO_2 sensor data, reducing computational complexity and dependency on multiple sensors. The experiment, conducted with data from Porto, employs Deep Learning (DL) techniques like MLP, LSTNet, Temporal Convolutional Networks (TCN), Transformers, and a Transformer-LSTM combination to predict NO_2 levels two days ahead.

The document is structured as follows: This section follows the case study, in which the dataset is described and analyzed, and the applied methods are presented. Then, section 3 details the experimentation and Section 4 includes the Results and Discussion. Finally, the conclusions and future works are exposed.

2 Material and Methods

This section begins with the case study, followed by the data analysis and preprocessing steps, and concludes with the DL models developed for forecasting NO_2 concentrations.

2.1 Case Study

The main goal of this research is to forecast NO_2 concentration in the city of Porto, Portugal. The dataset employed was retrieved from the OpenAQ platform [20], which aggregates open air quality data from official sources. Specifically, we used measurements recorded by the air quality monitoring station located on Fernando de Magalhães Street, in the Vigorosa neighbourhood (see Figure 1), with coordinates 41.16444° N, -8.58944° E.

This station, identified by the code NET-PT00015 in OpenAQ, is classified as an urban traffic station. It is situated in a densely populated area with significant road traffic, making it particularly suitable for studying pollution patterns related to vehicular emissions. The station provides hourly records of nitrogen dioxide (NO_2) concentration, measured in micrograms per cubic meter ($\mu g/m^3$), along with other pollutants and meteorological variables. Porto, being the second-largest city in Portugal, has faced increasing challenges related to air quality due to urban expansion and traffic congestion. The selected location is representative of typical European mid-sized urban areas and offers an insightful case for testing the robustness of Deep Learning (DL) models in real conditions.

Figure 1: Location or air monitoring station

2.2 Data Analysis

The dataset covers the period from May 10, 2018, to April 29, 2021, with a daily frequency. It comprises a time series that measures the mass of NO_2 in $\mu g/m^3$

and CO in $\mu g/m^3$, along with their corresponding dates, recorded in the format `YYYY-MM-DD`. The dataset consists of 1.086 recorded observations, among which 7 missing values were identified over the specified period. Table 1 presents all the dataset's features.

Table 1: Features of the collected dataset.

#	**Features**	**Description**	**Unit**	**Type**
1	*timestep*	Date of the Measurement	date	datetime
2	*co_value*	CO Mass Concentration	$\mu g/m^3$	float
3	*value*	NO_2 Mass Concentration	$\mu g/m^3$	float

The second data analysis phase involved a statistical analysis focused on the study's target, the NO_2 concentration. Table 2 displays the resulting statistical data.

Table 2: Descriptive statistics for target feature

Num. of items	**Min**	**Max**	**Mean**	**Median**	**Std. Deviation**	**Skewness**	**Kurtosis**
1086	7.50	102.81	48.29	44.53	21.49	0.64	-0.23

The table indicates that the average NO_2 concentration during the recorded period is 48.29 $\mu g/m^3$, slightly higher than the median value of 44.53 $\mu g/m^3$, suggesting a minor right skew due to higher concentration measurements. The standard deviation of 21.49 $\mu g/m^3$ reflects moderate variability, while the skewness value of 0.64 confirms a slight right skew, with higher concentrations more frequent than very low ones [21]. The kurtosis value of -0.23 indicates a platykurtic distribution with fewer extreme values than a normal distribution [21], showing that while variability exists, outliers have minimal impact. The NO_2 data exhibits moderate spread, a slight skew toward higher values, and a stable distribution without significant deviations.

After completing the statistical analysis, understanding the temporal variation of the target concentration value became a priority. As a result, two visualizations were created: one illustrating the daily variations in NO_2 concentration and another analyzing seasonal patterns over multiple years.

The line chart in Figure 2 presents the daily NO_2 concentrations recorded from 2018 to early 2021, highlighting fluctuations marked by pronounced peaks. The trendline underscores recurrent end-of-year peaks, indicating the influence of seasonal patterns. A significant increase in NO_2 levels is observed from November 2020 to February 2021, likely due to seasonal factors, such as intensified heating usage [22]. By 2021, a substantial decline in NO_2 concentrations, dropping below 20

$\mu g/m^3$, is evident, reflecting reduced industrial activities caused by the COVID-19 pandemic [23].

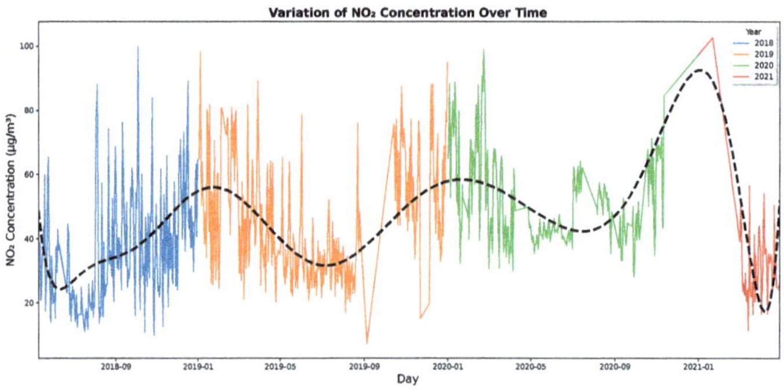

Figure 2: Daily NO_2 concentration variation over the years.

An examination of the average NO_2 concentration across different seasons from 2018 to 2021, as shown in Figure 3, reveals distinct seasonal patterns. NO_2 concentrations are consistently highest during winter, reaching a peak of about 80 $\mu g/m^3$ in 2021, most likely due to increased emissions from heating sources and meteorological conditions that limit atmospheric dispersion [24]. In contrast, summer has the lowest concentrations, reaching as low as 30 $\mu g/m^3$ in 2018. With average concentrations ranging from 30 to 60 $\mu g/m^3$, spring and autumn represent transitional periods impacted by a mix of advantageous and unfavourable dispersal circumstances. Understanding how seasonal changes affect pollution levels is crucial, and this comparative analysis highlights the necessity of strategic monitoring and year-round pollution control.

Consequently, this study employed the Kolmogorov-Smirnov (K-S) test to assess the Gaussian distribution of the dataset. This non-parametric test is commonly used to compare a sample with a reference probability distribution, such as the normal distribution. The K-S test was selected for its general applicability, as it does not rely on specific assumptions about the underlying distribution. A significance level of $p < 0.05$ was set to indicate that if the p-value falls below this threshold, the null hypothesis (which states that the data follows a Gaussian distribution) should be rejected. This analysis indicated that none of the attributes adhered to a Gaussian distribution, confirming significant deviations from normality.

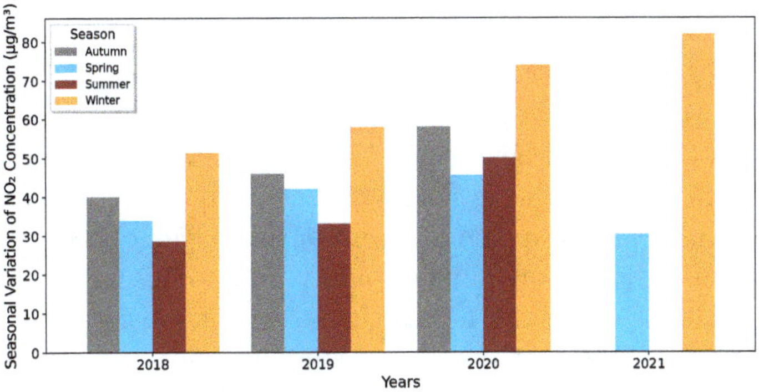

Figure 3: Average NO_2 concentration per season over the years.

2.3 Data Preprocessing

Upon completing the data analysis phase, preparing the dataset to feed the candidate models was imperative. The first task in data preprocessing involved addressing missing values identified during the exploratory analysis. The linear interpolation technique was chosen for its ability to continuously fill gaps while maintaining the natural timeline of the data [25].

Subsequently, feature engineering was performed to extract time-based attributes (e.g., month, day, year) from the original date-time feature, aiming to improve the models with enhanced temporal context. Spearman's correlation analysis, chosen for the non-normal data, revealed that only CO had a meaningful correlation with NO_2 (correlation coefficient: 0.6), while other variables, including the new features, showed correlations below 0.3. As a result, the dataset was refined to include only CO and NO_2 for model training.

The final data preparation stage involved converting the dataset into a format suitable for supervised learning, structuring the time series data into lagged feature sets, and ensuring that temporal dependencies were properly captured for model training. The MinMaxScaler method was applied to normalize the input features. For hybrid models utilizing LSTM, the data was normalized to a range of -1 to 1, whereas for the other models, it was normalized to a range of 0 to 1.

2.4 Deep Learning Models Conception

This section focuses on conceptualising DL models tailored to the preprocessed data. It explores how the models are built and highlights their relevance to the problem domain.

2.4.1 MLP

The MLP is one of the fundamental architectures in DL, known for its simplicity and effectiveness in solving a wide range of tasks [26]. It is a feedforward model that traditionally uses an input, hidden, and output layer. The first layer receives the input and flattens it to a one-dimensional vector to ensure the following layers can process the data. As for the hidden layer, which consists of a series of fully connected dense layers, the first one learns simple trends and relationships in the data, while deeper ones learn more complex details. To promote generalization and prevent overfitting, dropout is used throughout the network to turn off some neurons randomly. At the end, a dense layer with a single unit is outputted for regression tasks, which provides the predicted value. For a more general understanding, the following formula was developed to showcase the architecture of the MLP model:

$$\hat{\mathbf{y}} = \mathbf{W}_{out} \cdot \text{Dropout}(\text{HiddenLayers}(\text{Flatten}(\mathbf{X}))) + \mathbf{b}_{out} \tag{1}$$

and

$$\text{HiddenLayers}(F) = \sigma\left(\mathbf{W}_L \cdot \sigma\left(\ldots \sigma(\mathbf{W}_1 \cdot F + \mathbf{b}_1) \cdots + \mathbf{b}_L\right)\right) \tag{2}$$

where $\hat{\mathbf{y}}$ is the output, the formula begins by flattening the input $\mathbf{X}$. The `HiddenLayers` function then applies a series of dense layers with weights $\mathbf{W}_i$, biases $\mathbf{b}_i$, and activation functions σ to learn complex features from the input progressively. Afterwards, dropout is applied, with a rate $\mathbf{D}$. Finally, the output is passed through a final dense layer with weights $\mathbf{W}_{out}$ and bias $\mathbf{b}_{out}$ to produce the prediction $\hat{\mathbf{y}}$ [27].

2.4.2 LSTNet

The subsequent model used in this study is LSTNet, mainly used in Multivariate Time Series (MVTS) [28]. Our model's architecture is composed of Convolutional and LSTM layers. The convolutional layers capture local patterns and short-term dependencies, extracting high-level features, while the LSTM layers focus on long-term trends and temporal relationships that use information from earlier timesteps. Afterwards, the output from the previous layers passes through a skipRNN, which accounts for periodic patterns, by skipping specific time intervals. An Autoregressive (AR) component is then employed to detect trends quickly, passing to fully connected layers to create the final forecast. For the illustration of the architecture of the LSTNet model, the following formula was produced:

$$\hat{\mathbf{y}} = \text{output}\left(f\left(\mathbf{W}_{\text{out}} \cdot \text{dense}\left(\text{AR}\left(\text{skipRNN}\left(\text{LSTM}\left(\text{conv}(\mathbf{X})\right)\right)\right)\right) + \mathbf{b}_{\text{out}}\right)\right) \tag{3}$$

where $\hat{\mathbf{y}}$ is the final output, the formula applies a convolution on input $\mathbf{X}$, followed by LSTM, skipRNN, and AR layers, all processed through a dense layer. The result is scaled by the weights $\mathbf{W}_{\text{out}}$, passed through an output function f, and adjusted by the bias $\mathbf{b}_{\text{out}}$ to produce the prediction [29].

2.4.3 TCN

The succeeding model, TCN, is specifically designed to handle time series data. It uses causal convolutions, dilated convolutions and a fully-connected layer [30, 31]. Our model receives the input data and then passes it to the TCN layers, where it applies a causal convolution, where the prediction only depends on previous values in the data. It is then passed as the input for the dilated convolution layers. Additionally, dilations capture patterns over larger intervals without modifying the parameters, allowing the model to expand the receptive field. Ultimately, the features serve as the input in a densely connected layer with a softmax activation function for the prediction task. The formula for the TCN model was constructed to highlight the use of temporal convolutional networks:

$$\hat{\mathbf{y}} = f_{\text{softmax}}(\mathbf{W}_{\text{dense}} \cdot f_{\text{dilated}}(f_{\text{causal}}(\mathbf{X} * \mathbf{W}_{\text{causal}} + \mathbf{b}_{\text{causal}})) + \mathbf{b}_{\text{dense}}) \tag{4}$$

where the input $\mathbf{X}$ first passes through causal convolution layers, applying weights $\mathbf{W}_{\text{causal}}$ and adding bias $\mathbf{b}_{\text{causal}}$, followed by the application of a causal activation function f_{causal}, the result is then processed by a dilated activation function f_{dilated}. The output is further passed through a fully connected dense layer, where it is multiplied by dense weights $\mathbf{W}_{\text{dense}}$, and bias $\mathbf{b}_{\text{dense}}$ is added. Finally, a softmax activation function f_{softmax} is applied to produce the final prediction $\hat{\mathbf{y}}$ [32].

2.4.4 Transformers

A neural network architecture that uses attention mechanisms to process sequential data and capture relationships between elements is known as Transformers [33]. The input is first passed through a normalization layer, which adjusts and scales it to enhance the model's ability to learn effectively. It then is passed to a Multi-Head Attention (MHA) mechanism, which aims to focus on multiple parts of the input data and learn different relationships across time. The output from the previous step is forwarded to the feed-forward network, which transforms the feature space and creates a residual connection to the original input. Thereafter, a global average pooling is performed to reduce the dimensionality of the data, followed by MLP layers, to allow the model to learn more intricate information about the data and dropout is applied. At the end, the output from the previous steps is then passed to fully connected layers for prediction purposes. For the Transformer model, the formula was designed to represent the overall architecture:

$$\hat{\mathbf{y}} = \text{Dense}\left(\text{MLP}\left(\text{GlobalAveragePooling1D}\left(\text{MHA}\left(\text{LayerNorm}(\mathbf{X})\right)\right)\right)\right) \tag{5}$$

where $\mathbf{X}$ is the input, the model applies layer normalization (LayerNorm) followed by a MHA, then uses GlobalAveragePooling1D to aggregate the features. These features are further refined through MLP layers, producing the predictions $\hat{\mathbf{y}}$ [33].

2.4.5 Transformers-LSTM

The fifth and final model used is a hybrid approach of Transformers and LSTM, where we combine the feature extraction from the transformer and pass it to the LSTM to capture long-short term dependencies [34]. Our hybrid model uses the same approach as in section 2.4.4 by including the MLP layers part to have a richer feature extraction process. Consequently, we use the same construction method as in section 2.4.2, excluding the Convolutional part. The architecture of the Transformers-LSTM model is encapsulated in the following formula:

$$\hat{\mathbf{y}} = \sigma\left(\mathbf{W}_{\text{out}} \cdot \text{LSTM}\left(\text{Residual}\left(\text{Norm}(\mathbf{X}) + \text{MHA} + \text{FFN}\right)\right)\right) + \mathbf{b}_{\text{out}} \tag{6}$$

where $\mathbf{X}$ is the input sequence, the model applies Layer Normalization Norm(), MHA, and a Feed-Forward Network (FFN), followed by a Residual connection. The result is passed through an LSTM layer to capture sequential dependencies, and a Dense layer with a sigmoid activation, σ, generates the prediction $\hat{\mathbf{y}}$, using weights $\mathbf{W}_{\text{out}}$ and bias $\mathbf{b}_{\text{out}}$ [34].

3 Experiments

The experimental setup, including the methodology and hyperparameter optimization, is detailed in this section. Various experiments were conducted for the five DL models to identify the best combination for optimal model performance. Table 3 summarizes the hyperparameters considered and their respective values.

The hyperparameter search encompassed key factors influencing model performance. The number of timesteps was a primary focus, defined as the number of

Table 3: Hyperparameters' searching space.

Parameter	MLP	TCN	LSTNet	Transformers Transformers-LSTM
Layers	[3, 4]	[3, 4]	[3, 4]	[2, 4]
Neurons	[32, 64, 128]		[32, 64]	[32, 64, 128]
Activation	[ReLU, Tanh]	[ReLU, Tanh]	[ReLU, Tanh]	[ReLU, Tanh]
Timesteps	[7, 14, 21]	[7, 14, 21]	[7, 14]	[7, 14, 21]
Batch Size	[5, 10]	[10, 20]	[5, 10]	[5, 10]
Dropout	[0.0, 0.5]	[0.0, 0.5]	[0.0, 0.5]	[0.0, 0.5]
Kernel Size	-	[4, 5]	[2, 3]	[3, 4, 5]
Pool Size	-	[2, 3]	[2, 3]	[2, 3]
N.º of MLP Layers	-	-	-	[2, 3]
N.º of Heads	-	-	-	[4, 8]
Head Size	-	-	-	[128, 256]
Feed-Forward Dim	-	-	-	[256, 512]
Dilations	-	[[1,2,4], [1,2,4,8]]	-	-
N.º of Stacks	-	[2, 3]	-	-
Padding	-	[causal, same]	-	-
RNN Units	-	-	[1, 2]	-
Skip RNN Units	-	-	[1, 2]	-
Skip	-	-	[10, 50]	-
Filters	-	-	[16, 32]	-

past observations used to predict the next two days of NO_2 levels. Specifically, 7, 14, and 21 values were tested, representing one, two, and three weeks of historical data, respectively. Another key hyperparameter, the kernel size, defines the number of consecutive data points considered during convolution, improving the model's capacity for pattern recognition. [35]. Similarly, the dilation rate was examined, as it controls the spacing between filter elements, enabling the network to capture long-term dependencies while keeping computational costs manageable [36]. Lastly, in transformer models, the number of attention heads was analyzed for its effect on the model's ability to attend to multiple parts of the input sequence concurrently, improving the learned representations [37].

All experiments were evaluated using two different error metrics, the Root Mean Squared Error (RMSE) and Mean Absolute Error (MAE), ensuring consistent performance assessment. To mitigate reproducibility issues, a fixed seed value of 91195003 was applied across all experiments. Furthermore, overfitting and underfitting were addressed by generating and analyzing learning curves for each model to determine the optimal number of training epochs.

Given the time series nature of the problem, the *TimeSeriesSplit* cross-validation technique, with a k value of 3, was employed to preserve the chronological order of the data throughout the training and validation phases. Additionally, recursive forecasting was applied to predict future values iteratively, using prior predictions as inputs for subsequent forecasts. The experiments were conducted using Python

3.8, with essential libraries such as *numpy* and *pandas* for data analysis and pre-processing. The DL models were implemented using the *TensorFlow* v2.0.0 library. Google Colaboratory served as the platform for executing the experiments, providing adequate computational resources for efficient model training and evaluation.

4 Results and Discussion

The DL models, discussed in 2.4, were evaluated through a series of experiments, with the top 5 configurations for each model selected based on their RMSE performance. Table 4 and Table 5 details these configurations, including the corresponding hyperparameters for each entry.

Table 4: Best top-five results, per MLP, LSTNet and TCN models with all hyperparameters. The letters stand as follows: a. timesteps; b. batch size; c. n.º of layers; d. n.º of neurons; e. pool size; f. kernel size; g. dropout; h. activation; i. dilations; j. n.º of stacks; k. RNN units; l. skip RNN units; m. skip; n. padding; o. n.º of filters; p. RMSE; q. MAE.

a.	b.	c.	d.	e.	f.	g.	h.	i.	j.	k.	l.	m.	n.	o.	p.	q.
							MLP candidate models									
21	20	5	128	-	-	0.5	tanh	-	-	-	-	-	-	-	7.65	7.13
21	10	5	32	-	-	0.5	tanh	-	-	-	-	-	-	-	7.70	7.04
21	20	5	64	-	-	0.5	tanh	-	-	-	-	-	-	-	7.70	7.22
21	20	5	32	-	-	0.5	tanh	-	-	-	-	-	-	-	7.89	7.27
21	20	3	32	-	-	0.5	tanh	-	-	-	-	-	-	-	8.21	7.69
							LSTNet candidate models									
7	10	3	32	-	3	0.0	relu	-	-	1	1	10	-	16	8.68	8.14
7	10	3	64	-	2	0.0	relu	-	-	1	1	50	-	32	8.95	7.89
7	5	4	32	-	3	0.0	relu	-	-	1	2	10	-	16	9.32	8.54
7	10	4	32	-	2	0.0	relu	-	-	1	2	10	-	32	9.55	8.45
14	10	4	32	-	3	0.0	tanh	-	-	1	1	10	-	16	9.65	9.21
							TCN candidate models									
21	20	3	-	-	4	0.0	relu	[1, 2, 4, 8]	2	-	-	-	same	16	11.02	10.67
14	20	3	-	-	4	0.0	relu	[1, 2, 4, 8]	2	-	-	-	same	16	11.53	11.09
21	20	4	-	-	5	0.0	relu	[1, 2, 4, 8]	2	-	-	-	same	16	13.03	12.73
7	20	4	-	-	4	0.0	tanh	[1, 2, 4, 8]	2	-	-	-	same	16	13.03	12.71
7	20	4	-	-	5	0.0	relu	[1, 2, 4, 8]	2	-	-	-	casual	16	13.08	12.59

The MLP models performed the best, with an RMSE of 7.65 $\mu g/m^3$ and an MAE of 7.13 $\mu g/m^3$. Key hyperparameters across the top five models included timesteps (21), dropout (0.5), and activation function (tanh). The top three models showed similar RMSEs, with the first having the highest neurons (128), while the second used a smaller batch size (10) and fewer neurons (32). The worst-performing model had fewer layers (3), suggesting that more neurons and layers enhance performance by capturing complex patterns.

The top Transformer model ranked second with RMSE and MAE values of 8.13

Table 5: Best top-five results, per Tranformers and Transformers-LSTM models with all hyperparameters. The letters stand as follows: a. timesteps; b. batch size; c. n.º of layers; d. n.º of neurons; e. dropout; f. activation; g. n.º of encoding layers; h. n.º of MLP layers; i. n.º of heads; j. head size; k. feed-forward dimension; l. RMSE; m. MAE.

a.	b.	c.	d.	e.	f.	g.	h.	i.	j.	k.	l.	
Transformer candidate models												
21	5	4	128	0.0	relu	4	2	4	128	512	8.13	7.82
21	5	2	64	0.5	tanh	2	3	4	128	512	8.46	8.05
21	5	4	32	0.0	relu	4	3	4	128	512	8.85	8.41
21	5	4	64	0.0	relu	4	2	4	128	512	8.92	8.41
21	5	4	32	0.0	tanh	4	3	4	128	512	8.93	8.47
Transformer-LSTM candidate models												
21	10	2	64	0.0	relu	2	2	4	128	512	8.17	7.65
21	10	2	32	0.0	relu	2	2	4	128	512	8.25	7.86
21	10	2	64	0.5	tanh	2	3	4	128	512	9.01	8.35
21	10	2	64	0.5	relu	2	2	4	128	512	9.05	8.70
7	10	2	32	0.0	relu	2	2	4	128	512	9.05	8.44

$\mu g/m^3$ and 7.82 $\mu g/m^3$, respectively. Like MLP models, it used 128 neurons, while subsequent experiments used 64 and 32. The second experiment featured a distinct configuration with fewer layers (2), higher dropout (0.5), and fewer encoding layers (2). Despite these variations, timesteps (21), batch size (5), number of heads (4), head size (128), and feed-forward dimension (512) remained consistent across the top models.

Ranking third overall, the Transformer-LSTM achieved an RMSE of 8.17 $\mu g/m^3$ and a MAE of 7.65 $\mu g/m^3$. The fifth model, using only 7 timesteps instead of 21, matched the fourth model's RMSE of 9.05 $\mu g/m^3$ despite employing half as many neurons (32). Experiments three and four featured a higher dropout rate (0.5), with the third also using a different activation function and more MLP layers, achieving an RMSE of 9.01 $\mu g/m^3$. Other key parameters—encoding layers (2), number of heads (4), head size (128), feed-forward dimension (512), and batch size (10)—remained uniform across all models.

The fourth-ranked model, LSTNet, achieved a best RMSE of 8.68 $\mu g/m^3$. The first four models used a timestep of 7 and the ReLU activation function, in contrast to the last candidate model, which used a timestep of 14 and the tanh activation function. The third model employed a smaller batch size of 5, resulting in a RMSE of 9.32 $\mu g/m^3$, while the second model doubled the number of neurons to 64 and used a higher skip value of 50. All models consistently shared a dropout rate 0.0

and a single RNN unit.

The TCN models delivered the weakest performance, with the best RMSE at 11.02 $\mu g/m^3$, nearly 4 $\mu g/m^3$ worse than the MLP models. The top three experiments used higher timesteps (21 for the first and third, 14 for the second), while others used 7. Most hyperparameters were consistent, including dilations [1, 2, 4, 8], 16 filters, and a batch size 20. However, the fourth model used tanh activation, and the last model applied *causal* padding. Layer differences, 3 in the first two models and 4 in others influenced RMSE variability, with most models achieving 1.5–2 points higher than the best result.

Among all models, the MLP achieved the best performance with a top RMSE of 7.65 $\mu g/m^3$, thanks to consistent hyperparameters and a higher number of neurons and layers in the top-performing configuration. Transformer models ranked second with a best RMSE of 8.13 $\mu g/m^3$, supported by stable batch size, number of heads, and feed-forward dimensions, though variations in activation functions and dropout rates affected performance. Close behind, the Transformer-LSTM models reached a best RMSE of 8.17 $\mu g/m^3$, with differences in timesteps, neurons, and activation functions impacting results. LSTNet models followed in fourth place, achieving a best RMSE of 8.68 $\mu g/m^3$, influenced by variations in layers, kernel size, batch size, and skip values. Finally, the TCN models showed the weakest performance, with the best RMSE of 11.02 $\mu g/m^3$, limited by consistently higher RMSE values across all configurations despite differences in padding, activation functions, and timesteps. Overall, MLP models outperformed the rest, while Transformer-based architectures showed potential for further refinement.

5 Conclusion and Future Work

Understanding and addressing air quality challenges is essential for safeguarding public health and environmental sustainability. NO_2 levels are a critical indicator of air quality, with significant implications for public and planetary health. Elevated concentrations are often associated with urbanization, industrial activity, and vehicular emissions. Additionally, predicting it can be challenging due to meteorological factors and temporal variations, such as seasonal and diurnal cycles. Therefore, this study focused on forecasting the next two days of NO_2 concentrations using five DL models: MLP, LSTNet, TCN, Transformers and Transformers-LSTM.

Several experiments were conducted using a diverse range of hyperparameter configurations, yielding an unanticipated result: the MLP, a relatively simple model compared to its more complex counterparts. It has emerged as the best-performing model, with a RMSE value of 7.65 $\mu g/m^3$ and a MAE value of 7.13 $\mu g/m^3$. The other models, LSTNet, Transformer and Transformer-LSTM, achieved RMSE values between 8.0 $\mu g/m^3$ and 10.0 $\mu g/m^3$. In contrast, the TCN models were last with a RMSE score between 11.0 $\mu g/m^3$ and 13.1 $\mu g/m^3$.

Consequently, this study is a work that can be further explored in future investigations, exploring additional features correlated with NO_2, such as NO or Ozone, which may help reduce forecasting errors because this study only used one variable correlated with the target. Specialized models like LSTNet and TCN could benefit from larger datasets, enabling them to capture complex trends more effectively. Furthermore, enhancing the hyperparameter searching space offers another avenue for refining the model's performance. Finally, integrating alternative DL models or hybrid approaches, such as BI-LSTM, could further improve forecasting performance by leveraging the strengths of different architectures.

Acknowledgments

This work is financed by National Funds through the Portuguese funding agency, FCT - Fundação para a Ciência e a Tecnologia within project 2022.068 22.PTDC (`https://doi.org/10.54499/2022.06822.PTDC`). The work of Pedro Oliveira was supported by the doctoral Grant PRT/BD/154311/2022 financed by the Portuguese Foundation for Science and Technology (FCT), and with funds from European Union, under MIT Portugal Program.

CITIC, as a center accredited for excellence within the Galician University System and a member of the CIGUS Network, receives subsidies from the Department of Education, Science, Universities, and Vocational Training of the Xunta de Galicia. Additionally, it is co-financed by the EU through the FEDER Galicia 2021-27 operational program (Ref. ED431G 2023/01)

Xunta de Galicia. Grants for the consolidation and structuring of competitive research units, GPC (ED431B 2023/49).

This research is the result of the Strategic Project "Critical infrastructures cybersecure through intelligent modeling of attacks, vulnerabilities and increased security of their IoT devices for the water supply sector" (C061/23), as a result of the collaboration agreement signed between the National Institute of Cybersecurity (INCIBE) and the University of A Coruña. This initiative is carried out within the framework of the funds of the Recovery Plan, Transformation and Resilience Plan funds, financed by the European Union (Next Generation).

References

[1] N. I. of Corrections, Worldometers - world population clock (2023).

[2] D. Mintzia, F. Kehagia, A. Tsakalidis, E. Zervas, A methodological framework for the comparative analysis of the environmental performance of roadway and railway transport, PROMET - Traffic&Transportation 30 (6) (2018) 721–731. doi:10.7307/ptt.v30i6.2804.

[3] M. A. Khan, M. Z. Khan, K. Zaman, L. Naz, Global estimates of energy consumption and greenhouse gas emissions, Renewable and Sustainable Energy Reviews 29 (2014) 336–344. doi:10.1016/j.rser.2013.08.091.

[4] A. Ghosh, Possibilities and challenges for the inclusion of the electric vehicle (ev) to reduce the carbon footprint in the transport sector: A review, Energies 13 (10) (2020) 2602. doi:10.3390/en13102602.

[5] A. T. Hoang, et al., Power generation characteristics of a thermoelectric modules-based power generator assisted by fishbone-shaped fins: Part ii–effects of cooling water parameters, Energy Sources, Part A: Recovery, Utilization, and Environmental Effects 43 (3) (2021) 381–393. doi:10.1080/15567036.2019.1624891.

[6] D. N. Cao, et al., Effects of injection pressure on the nox and pm emission control of diesel engine: A review under the aspect of pcci combustion condition, Energy Sources, Part A: Recovery, Utilization, and Environmental Effects 46 (1) (2024) 7414–7431. doi:10.1080/15567036.2020.1754531.

[7] B. Kamarehie, et al., Quantification of health effects related to so2 and no2 pollutants by using air quality model, Journal of Advances in Environmental Health Research 5 (1) (2017) 44–50.

[8] M. B. Marinov, et al., Air quality monitoring in urban environments, in: 2016 39th International Spring Seminar on Electronics Technology (ISSE), 2016, pp. 443–448. doi:10.1109/ISSE.2016.7563237.

[9] Y. Yao, et al., Ambient nitrogen dioxide pollution and spreadability of covid-19 in chinese cities, Ecotoxicology and environmental safety 208 (2021) 111421. doi:10.1016/j.ecoenv.2020.111421.

[10] M. R. Miller, D. E. Newby, Air pollution and cardiovascular disease: car sick, Cardiovascular Research 116 (2) (2019) 279–294. doi:10.1093/cvr/cvz228.

[11] M. Hadei, et al., Estimation of mortality and hospital admissions attributed to criteria air pollutants in tehran metropolis, iran (2013-2016), Aerosol and air quality research 17 (10) (2017) 2474–2481.

[12] N. Karna, D. L. Lubna, S. Y. Shin, Air quality measurement device using programmable quadcopter drone towards internet of drone things, in: 2021 International Conference on Information and Communication Technology Convergence (ICTC), IEEE, 2021, pp. 753–758.

[13] O. Tayan, Context-aware framework for enhanced smart urban pollution monitoring and control, in: 2022 International Conference on Emerging Trends in Computing and Engineering Applications (ETCEA), 2022, pp. 1–5. doi:10.1109/ETCEA57049.2022.

10009678.

[14] A. Heydari, et al., Air pollution forecasting application based on deep learning model and optimization algorithm, Clean Technologies and Environmental Policy (2022) 1–15doi:10.1007/s10098-021-02080-5.

[15] M. Korunoski, B. R. Stojkoska, K. Trivodaliev, Internet of things solution for intelligent air pollution prediction and visualization, in: IEEE EUROCON 2019 -18th International Conference on Smart Technologies, 2019, pp. 1–6. doi:10.1109/EUROCON.2019.8861609.

[16] H. Hajmohammadi, B. Heydecker, Multivariate time series modelling for urban air quality, Urban Climate 37 (2021) 100834, https://discovery.ucl.ac.uk/10125569/1/VARMA doi:10.1016/j.uclim.2021.100834.

[17] J. A. Kamińska, Probabilistic forecasting of nitrogen dioxide concentrations at an urban road intersection, Sustainability 10 (11) (2018) 4213. doi:10.3390/su10114213.

[18] S. R. Shams, et al., Artificial intelligence accuracy assessment in no2 concentration forecasting of metropolises air, Scientific Reports 11 (1) (2021) 1805. doi:10.1038/s41598-021-81455-6.

[19] S. Agarwal, et al., Air quality forecasting using artificial neural networks with real time dynamic error correction in highly polluted regions, Science of the Total Environment 735 (2020) 139454. doi:10.1016/j.scitotenv.2020.139454.

[20] OpenAQ, Net-pt00015, https://explore.openaq.org/locations/7571 [Accessed: (01/20/2024)] (2024).

[21] G. Hatem, J. Zeidan, M. Goossens, C. Moreira, Normality testing methods and the importance of skewness and kurtosis in statistical analysis, BAU Journal-Science and Technology 3 (2) (2022) 7. doi:10.54729/KTPE9512.

[22] A. Siddiqui, et al., Deciphering seasonal variability and source dynamics of urban pollutants over delhi under surface meteorological influence using ground-based and trajectory modeling techniques, Earth Systems and Environment (2024) 1–17doi:10.1007/s41748-024-00541-7.

[23] V. Fioletov, et al., Quantifying urban, industrial, and background changes in no_2 during the covid-19 lockdown period based on tropomi satellite observations, Atmospheric Chemistry and Physics 22 (6) (2022) 4201–4236. doi:10.5194/acp-22-4201-2022.

[24] A. Andriulė, E. Vasiliauskienė, P. Rapalis, I. Dailidienė, Air pollution in the port city of lithuania: Characteristics of the distribution of nitrogen dioxide and solid particles when assessing the demographic distribution of the population, Sustainability 16 (19) (2024). doi:10.3390/su16198413.

[25] Y. Yang, et al., A study on water quality prediction by a hybrid cnn-lstm model with attention mechanism, Environmental Science and Pollution Research 28 (39) (2021) 55129–55139. doi:10.1007/s11356-021-14687-8.

[26] G. R. MR, N. Somu, A. P. Mathur, A multilayer perceptron model for anomaly detection in water treatment plants, International Journal of Critical Infrastructure Protection 31 (2020) 100393. doi:10.1016/j.ijcip.2020.100393.

[27] J. Isabona, et al., Development of a multilayer perceptron neural network for optimal predictive modeling in urban microcellular radio environments, Applied Sciences 12 (11) (2022). doi:10.3390/app12115713.

[28] W. Hong, M. Chen, P. Gao, D. Hong, Medium- and long-term orbit prediction of satellite based on lstnet, in: 2023 15th International Conference on Computer and Automation Engineering (ICCAE), 2023, pp. 534–540. doi:10.1109/ICCAE56788.2023.10111400.

[29] R. Liu, L. Chen, W. Hu, Q. Huang, Short-term load forecasting based on lstnet in power system, International Transactions on Electrical Energy Systems 31 (12) (2021) e13164. doi:10.1002/2050-7038.13164.

[30] J. Bi, X. Zhang, H. Yuan, J. Zhang, M. Zhou, A hybrid prediction method for realistic network traffic with temporal convolutional network and lstm, IEEE Transactions on Automation Science and Engineering 19 (3) (2022) 1869–1879. doi:10.1109/TASE.2021.3077537.

[31] R. Zhu, W. Liao, Y. Wang, Short-term prediction for wind power based on temporal convolutional network, Energy Reports 6 (2020) 424–429, 2020 The 7th International Conference on Power and Energy Systems Engineering. doi:10.1016/j.egyr.2020.11.219.

[32] A. K. Shaikh, A. Nazir, N. Khalique, A. S. Shah, N. Adhikari, A new approach to seasonal energy consumption forecasting using temporal convolutional networks, Results in Engineering 19 (2023) 101296. doi:https://doi.org/10.1016/j.rineng.2023.101296.

[33] A. Katharopoulos, A. Vyas, N. Pappas, F. Fleuret, Transformers are rnns: Fast autoregressive transformers with linear attention, in: International conference on machine learning, PMLR, 2020, pp. 5156–5165. doi:10.48550/arXiv.2006.16236.

[34] M. M. Hossain, et al., Transnet: Deep attentional hybrid transformer for arabic posts classification, IEEE Access 12 (2024) 111070–111096. doi:10.1109/ACCESS.2024.3441323.

[35] R. Rick, L. Berton, Energy forecasting model based on cnn-lstm-ae for many time series with unequal lengths, Engineering Applications of Artificial Intelligence 113 (2022) 104998. doi:10.1016/j.engappai.2022.104998.

[36] M. Wang, F. Qin, A tcn-linear hybrid model for chaotic time series forecasting, Entropy 26 (6) (2024) 467. doi:10.3390/e26060467.

[37] H. S. Oliveira, H. P. Oliveira, Transformers for energy forecast, Sensors 23 (15) (2023) 6840. doi:10.3390/s23156840.

 Received 10 January 2025

Defect Detection in Batavia and Sarga Woven Fabrics by Means of Convolutional Neural Networks

Beatriz Gil Arroyo, Nuria Velasco-Pérez, Ángel Arroyo, Daniel Urda
Grupo de Inteligencia Computacional Aplicada (GICAP), Departamento de Digitalización, Escuela Politécnica Superior, Universidad de Burgos, Spain
{bgarroyo,nuriavp,aarroyop,durda}@ubu.es

Juan Marcos Sanz
Textil Santanderina, Cabezón de la Sal, Spain
juanmarcos@tsanta.es

Abraham Casas
Centro Tecnológico de Componentes-CTC, Scientific and Technological Park of Cantabria (PCTCAN), Santander, Spain
acasas@centrotecnologicoctc.com

Abstract

Quality assurance stands as a pivotal phase across all manufacturing processes, particularly within the textile industry. Presently, textile inspections heavily rely on human visual assessment due to deficiencies in commercial solutions. In response, this research advocates for the adoption of Deep Learning models to automate fabric quality control within dynamic and contemporary production environments. Specifically, Convolutional Neural Networks are scrutinized using authentic images sourced from the Batavia and Sarga weave production lines. The experimental results identify DenseNet121 and InceptionV3 as the most effective models for Batavia and Sarga weaves, respectively. DenseNet121 demonstrates balanced performance across key metrics for Batavia weave, while InceptionV3 excels in Sarga weave, particularly in F1-score and AU-ROC. These findings underscore the potential of DL models to enhance the accuracy and efficiency of textile quality control.

Keywords: Textile defect detection; Fabric quality control; Industry 4.0; Deep Learning in manufacturing; Convolutional neural networks; Image analysis and classification.

1 Introduction and previous work

In recent years, various challenges in industrial sectors have been addressed through the application of soft computing methodologies [1]. Despite this progress, significant hurdles persist within the textile industry, which remains entrenched in traditional practices with limited technological integration. The sector's ability to withstand change can be attributed to its historical composition, primarily comprising long-standing entities adept at adapting to market shifts and competing effectively against low-cost alternatives. However, as the textile industry increasingly emphasizes competitiveness and elevates quality standards, the existing quality control mechanisms are proving inadequate and beset with limitations. Consequently, there arises an imperative to explore novel, automated, and intelligent inspection techniques.

Presently, quality inspection in the textile domain predominantly relies on visual scrutiny conducted by trained personnel, operating at an average pace of approximately 20 m/min, supplemented by sporadic quality audits. However, this inspection rate falls short of the manufacturing speed, which averages approximately 42 m/min, resulting in a bottleneck during the quality evaluation. While traditional visual inspection has served its purpose historically [2], its efficacy is waning, necessitating innovative solutions [3].

This study proposes and validates the efficacy of Deep Learning (DL) models in overseeing fabric quality at Textil Santanderina (TS) to address this challenge. The primary objective is to detect defects (including warp and weft irregularities, impurities, bowing, skewing, and color gradients) arising at various production stages (spinning, weaving, coating, and garment manufacturing). Specifically, Convolutional Neural Networks (CNNs) are advocated for their proven image processing capabilities in quality control applications [4]. Different CNN architectures will be evaluated to identify the most suitable model for the problem at hand. Despite the promise of artificial vision systems, particularly for inspection purposes, their industrial implementation remains constrained by persistent challenges, such as production interruptions following defect detection, delays, resource wastage, and over-reliance on human intervention. Although off-the-shelf image processing systems are available for textile applications, their lack of Artificial Intelligence (AI) integration hinders swift training and adaptation to the industry's high product variability (TS handles batches ranging from 500 m to 50,000 m, with over 700 new references annually).

The subsequent sections of this article are structured as follows: Firstly, a review of the relevant literature pertaining to the current investigation is presented in Section 2. Section 3 introduces the investigated issue and outlines the dataset generation process employed in this study. The methodological approaches utilized and the subsequent experimentation are delineated in Section 4. Section 5 provides a detailed exposition of the results of the experimentation. Finally, in Section 6, the best-performing models are identified, which also presents the main conclusions derived from the experimentation and the associated results.

2 Related work

Several commercial systems currently in use include Uster® EVS FABRIQ VISION, Mahlo GmbH, and PEKAT VISION, each with its own set of limitations. For instance, the Uster® EVS FABRIQ VISION system is restricted to post-production inspection and is suitable only for uniform production batches. Similarly, the Mahlo GmbH system is limited to detecting defects that do not affect the fabric's structure. At the same time, PEKAT VISION is primarily designed to easily identify visible errors in long batches, lacking an AI learning system for pattern recognition and series analysis. In recent years, various AI techniques have been used for the monitoring of quality and manufacturing processes in industry in general [5] and in the textile industry in particular [6]. Focusing on AI techniques for image analysis [7], it can be seen that the techniques selected in this research work are of current relevance due to their validity in this field. DL [8] architectures have been extensively utilized for quality control across various industries[9][10]. Specifically, CNNs [11] have demonstrated significant potential in manufacturing and quality assurance [12][13][14]. While some researchers have employed traditional methods [15] for defect detection in the textile sector, others have explored DL-based approaches. Notably, early studies such as [16] employed supervised CNNs, albeit on outdated datasets that fail to meet current industry standards. Other DL models, like those described in [17], have focused on detecting multiple defects within a single fabric image, which does not represent real-world scenarios where simultaneous defects are rare. Alternatively, some researchers have proposed unsupervised learning methods to streamline hyperparameter initialization and labeling efforts. For example, [18] introduced techniques to reduce the time and effort required for model fine-tuning, while [19] proposed the use of Variational Autoencoders to label defect-free images automatically. However, in our study, manual labeling by human operators is currently employed, making these approaches less relevant to our context. Overall, previous studies have primarily been conducted offline and not validated under real-world conditions with high manufacturing speeds.

The work presented at 18th International Conference on Soft Computing Models in Industrial and Environmental Applications (SOCO 2023) [20], demonstrated the feasibility of using DL for warp and weft defect detection. Although that study focused on the Batavia weave, this research delves into more fibers such as the Twill weave. Furthermore, a greater number of CNN pre-trained have been used in both tissues.

Therefore, our present research represents a significant advancement by validating CNNs under complex, high-speed production environments using up-to-date production conditions.

3 Case Study

In this study, CNNs are leveraged for analyzing images captured during the fabric production process at TS, a prominent European company specializing in yarns, fabrics, finishes, and garments made from various materials. Established in December 1960, TS boasts a rich history in the textile industry and currently offers a diverse range of products spanning the fashion/clothing, sportswear, healthcare, medical, military, public, and protective sectors.

Distinguished by its vertical supply chain, TS manages all aspects of fabric manufacturing, including spinning, weaving, coating finishing, garment production, and logistics. This comprehensive integration allows the company to deliver processed products at any stage of production with various finishes. With an annual output exceeding 40 million meters of fabrics and over 7 million garment units worldwide, TS is a key player in the textile market.

To support our research objectives, TS has equipped its facility with a Basler raL camera featuring the Awaiba DR-2k-7 CMOS sensor, capable of delivering 51 kHz at a resolution of 2k. Operating within the VIS-NIR bandwidth, the camera is equipped with Basler proprietary optics, enabling a resolution of 10 px/mm. For uniform illumination, a LED array emitting at 850nm, spanning a field of 15mm x 1510mm and comprising 360 infrared (IR) 850nm LEDs, is positioned 20 cm above the fabric at a 15º angle of incidence. This infrared (IR) illumination ensures compatibility with existing D65 standard illuminating systems for visual inspection. A depiction of the factory setup is provided in Figure 1.

The camera is synchronized with the fabric using a 10-bit inductive encoder to ensure high precision (<1/10 mm) positioning. CNNs are trained on the captured images following standard training, testing, and validation procedures outlined in Section 4. Lighting conditions, setup geometry, optics, and CMOS sensor specifications have been carefully chosen to meet additional requirements for defect inspection, including fabric width, batch length, and inspection speed, among others.

The experimentation has been conducted using two different weaves (Batavia and Sarga) in the research context. These textiles were selected due to their distinctive characteristics and relevance in various textile applications. The inclusion of both textiles in the study provides a comprehensive assessment of the efficacy of the proposed methods across a broader range of textile inspection scenarios. Additionally, it allows for comparing and contrasting the performance of machine learning algorithms under variable conditions, enriching the understanding of their ability to address specific challenges in the textile industry. To conduct this study, images provided by TS were utilized and subsequently compiled into a public dataset titled "Original and processed dataset of Batavia and Sarga woven fabric images" [21]. This dataset is further discussed in an article titled "Detailed description of the classified image dataset of a textile company in Spain"[22].

3.1 Batavia

For the Batavia fabric, it's important to note that Batavia is a specific type of twill weave, categorized under the broader classification of Sarga fabrics. Images of a Batavia fabric were gathered in November 2022 over several days. This fabric is a blend of cotton and linen (85/15) with a standard twill weave known as Batavia (refer to Figure 1). A neutral color (greige fabric) was selected for its widespread use in woven fabrics for the summer season and to minimize variability in color combinations and patterns. A total of 2,755 high-resolution grayscale images, each 16 bits, with dimensions of 2,048x696 pixels were collected (as illustrated in Figure 2).

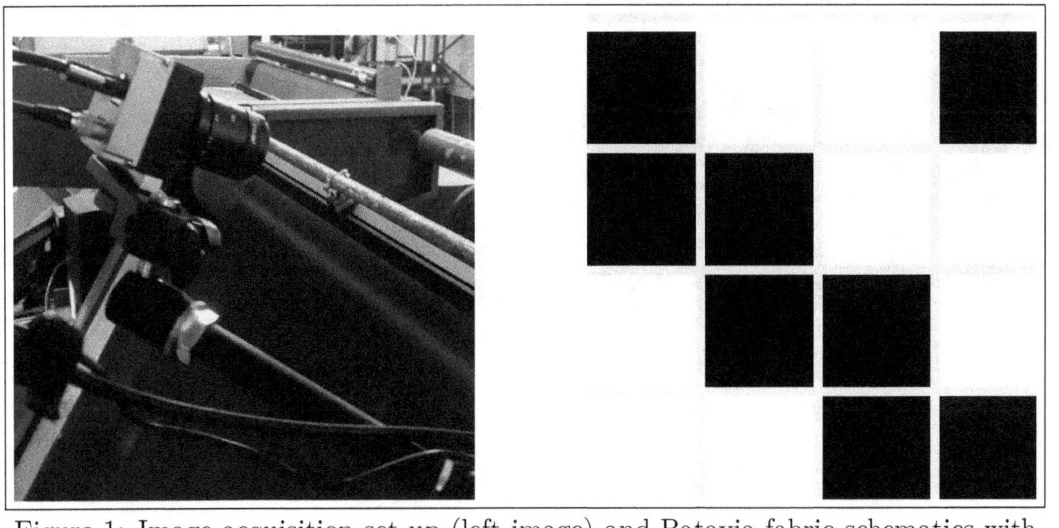

Figure 1: Image acquisition set up (left image) and Batavia fabric schematics with 4 yarns warp and 4 yarns weft (right image).

Figure 2: Example of 2,048X696 original Batavia image.

3.2 Sarga

The Sarga fabric, which is characterized by its distinctive diagonal weave pattern, offers durability and versatility, making it suitable for various textile applications. A neutral color (greige fabric) was selected to minimize variability in color combinations and patterns. A total of 1,548 high-resolution grayscale images, each 16 bits, with dimensions of 2,048x696 pixels were collected. In Figure 3 an image of Sarga fabric can be observed.

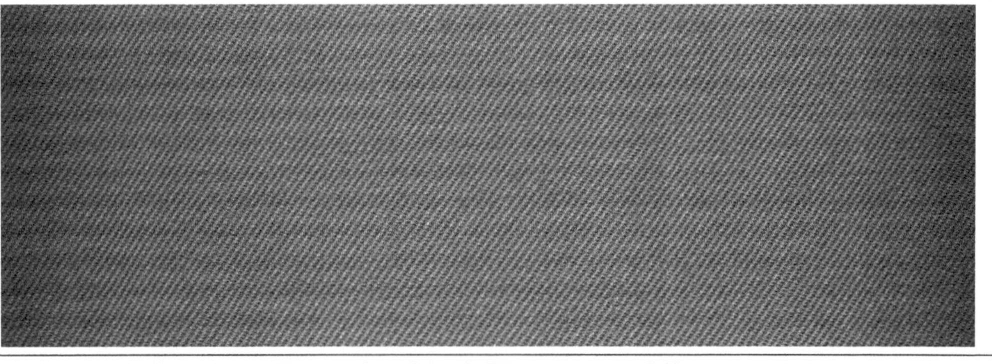

Figure 3: Example of 2,048X696 original Sarga image.

4 Methods and Experimentation

This section delineates the DL methodologies and strategies employed in this study to craft a precise model for fabric defect detection. Additionally, comprehensive insights into the experimental design are furnished within this section, with specific methodologies elaborated in subsection 4.1 and the experimentation outlined in subsection 4.2.

4.1 Methods

CNNs [23] have become increasingly recognized as a robust tool for tasks such as object detection and image analysis. As larger datasets have become more accessible, CNNs have demonstrated notable enhancements in performance, facilitating the creation of autonomous systems applicable across diverse domains. This study aims to harness the capabilities of CNNs to support quality control teams in promptly identifying fabric defects. Such assistance aims to enhance the quality of end products, lower operational expenses, and augment customer satisfaction.

To achieve this objective, the examination conducted in this study centers on the dataset delineated in the preceding section, wherein the gold standard entails the classification of fabric images meticulously evaluated by a quality control specialist. In this context, the authors suggest assessing the benefits of constructing two CNN models with analogous architectures applied to images within the specified dataset, along with employing transfer learning architectures to evaluate their efficacy compared to a standard labeling baseline.

Firstly, we are going to review the basis of CNNs and transfer learning. CNNs are a type of deep neural network that is designed to process data that has a grid-like structure (such as images) and use convolutional layers to extract relevant features from the image by applying a set of filters to the input. These filters can learn to detect specific patterns and structures in the images, such as edges, corners, and textures. CNNs typically include pooling layers, which down-sample the feature maps generated by the convolutional layers, as well as fully connected layers, which classify the features into specific categories. On the other hand, transfer learning is a DL technique in which a model of CNNs trained on one task is reused as a starting point for training a model on a different task. In other words, transfer learning involves using pre-trained neural network models as a starting point for training a new model on a smaller dataset or a different task. We can differentiate two types of transfer learning: in feature extraction, the pre-trained model is used as a fixed feature extractor, where only the last few layers of the model are modified to fit the new task and in fine-tuning, the pre-trained model is used as a starting point, but all of the layers are updated and retrained on the new task. In this study, we have focused on this last technique.

Next, the baseline and CNNs methods employed in this work are described.

4.1.1 Baseline

This entails the quality control team's real-time classification of products during the final stages of the production line. This performance benchmark aligns with the level of human accuracy we aim to achieve, or ideally surpass, through the utilization of DL-based methods for this intricate classification task.

The baseline methodology employed in this study does not necessitate the implementation of any specific model, as the dataset already contains labels assigned by the quality control team. Additionally, analogous labels are provided for the gold standard, derived offline by an expert who meticulously examines all images. Consequently, a performance metric can be readily computed to evaluate human proficiency in fabric defect detection.

4.1.2 Pre-trained CNNs

Popular pre-trained models have been adjusted for the given problem and fine-tuned on the datasets, under the "transfer learning" paradigm. The applied models are listed in chronological order of their introduction, reflecting the historical evolution of CNNs architectures:

- **VGG16**: It is a pre-trained CNN model that was developed by the Visual Geometry Group at the University of Oxford [24]. The model consists of 16 layers, including 13 convolutional layers and 3 fully connected layers, and was trained on the ImageNet dataset, which contains millions of labelled images. VGG16 is known for its simple and uniform architecture, which uses small 3x3 filters throughout the convolutional layers and max pooling layers to downsample the feature maps. The model has been shown to perform well on a wide range of image classification tasks, including object recognition and localization. One of the main advantages of VGG16 is its flexibility and ease of use.

- **VGG19**: It is a CNN model that also was developed by the Visual Geometry Group at the University of Oxford [24]. It is an extension of the VGG16 model and has 19 layers, including 16 convolutional layers and 3 fully connected layers. Like VGG16, VGG19 was trained on the ImageNet dataset. VGG19 is known for its deep architecture and uses a similar approach as VGG16, with small 3x3 filters and max pooling layers to downsample the feature maps. However, VGG19 has more layers and uses larger filters in the later layers, which allows it to capture more complex features in the input images.

- **InceptionV3**: It is a CNN model that was developed by Google researchers as part of the Inception architecture family [25]. The model consists of 48 convolutional layers, including traditional convolutional layers and Inception modules, which use multiple parallel convolutions of varied sizes to extract features from the input images. Some unique features of the network are that it uses batch normalization after every convolutional and fully connected layer to accelerate the training, and it uses factorization into smaller convolutions (such as 1x1 and 3x3 convolutions) in order to reduce the computational complexity of the model. InceptionV3 was trained on the ImageNet dataset with a top-5 error rate of less than 4%. The model is known for its efficiency and accuracy.

- **ResNet101**: It is a CNN model that was developed by researchers at Microsoft as part of the ResNet (Residual Network) family of architectures [26]. The model consists of 101 layers, with a total of over 44 million parameters. It makes use of residual connections to allow for the efficient training of very deep neural networks. These connections allow information from earlier layers of the network to be passed directly to later layers, bypassing the need to go through multiple non-linear transformations. In addition, ResNet101 uses a bottleneck architecture, which reduces the number of parameters in the model while increasing its depth. This is achieved by using 1x1 convolutions to reduce the number of channels in the input tensor before applying 3x3 convolutions. And, finally, ResNet101 uses pre-activation, which means that batch normalization and activation functions are applied before convolutions.

- **ResNet50V2**: It is a CNN architecture belonging to the ResNet family, designed to improve the efficient training of deep networks [27]. This model comprises 50 layers and leverages residual connections to mitigate the vanishing gradient problem, facilitating the training of very deep networks. ResNet50V2 adopts a pre-activation scheme, where batch normalization and activation functions are applied before convolutions in each residual block, enhancing gradient flow and training stability. Additionally, it employs a bottleneck architecture, combining 1x1, 3x3, and 1x1 convolutions to reduce the dimensionality of intermediate data while maintaining computational efficiency. This optimized design makes it particularly well-suited for complex computer vision tasks requiring both depth and efficiency.

- **Xception**: It is a CNN model that was developed by Google researchers in 2016 [28]. The model is based on the Inception architecture, which uses multiple parallel convolutions of different sizes to extract features from the input images. However, unlike Inception, Xception uses depthwise separable convolutions, which split the convolution operation into two separate operations: a depthwise convolution and a pointwise convolution. The use of depthwise separable convolutions in Xception allows the model to have a much smaller number of parameters than traditional CNNs. This makes Xception a computationally efficient model that is suitable for deployment on mobile devices and other resource-constrained environments.

- **DenseNet121**: It is a CNN model that was developed by researchers at Facebook AI Research [29]. The model consists of 121 layers and has a total of approximately 8 million parameters. One of the unique features of DenseNet121 is its use of dense connections, which allow information to be passed between all layers of the network. This helps to improve the flow of information through the network and makes it more efficient at learning complex representations of visual features. DenseNet121 also makes use of batch normalization and ReLu activation functions, which have been shown to improve the accuracy of the model. Additionally, the model makes use of a global average pooling layer, which reduces the number of parameters in the network and helps to prevent overfitting.

- **EfficientNetB0**: It is the baseline model in the EfficientNet family, comprising 5.3 million parameters [30]. It employs a compound scaling method that simultaneously adjusts the network's depth, width, and input resolution, optimizing both efficiency and performance. With an input resolution of 224x224, EfficientNetB0 achieves an optimal balance between accuracy and computational cost, making it particularly suitable for real-time computer vision tasks, such as image classification and object detection, in resource-constrained environments. This model outperforms traditional CNNs, such as ResNet50, with fewer parameters while maintaining competitive accuracy.

- **EfficientNetB3**: It extends the architecture of EfficientNetB0 by increasing the network's depth, width, and input resolution to 300x300, resulting in a model with 12 million parameters [30]. This enhancement allows EfficientNetB3 to capture more complex patterns and improve accuracy compared to the baseline model. The compound scaling approach ensures that the increase in model size does not lead to a disproportionate increase in computational cost. EfficientNetB3 is well-suited for applications that require higher accuracy, such as advanced image classification and object detection tasks, while maintaining computational efficiency relative to its performance.

Throughout the training procedure, specific hyperparameters were set as follows: a learning rate set to 0.002, 2 layers dense of neurons, with 8,192 neurons per layer, dropout per layer to 0, batches consisting of 128 images were used to train and update the parameters of DL models for up to 100 epochs, regulated by an early stopping system. Mechanism designed to stop training if no improvement in loss is observed on a validation set after 10 epochs.

4.2 Experimentation

The methodology employed in this investigation to develop and assess the efficacy of a CNN model for the specified quality control task encompasses several pivotal stages, which are delineated in further detail below.

4.2.1 Image Preprocessing

Initial image preprocessing was conducted wherein the original 16-bit images underwent rescaling to their corresponding 8-bit versions. This transformation reduced the potential pixel values per image from 65,536 to 256. Subsequently, to facilitate meticulous fault detection and expand the dataset, the 8-bit images were partitioned into twelve segments, each measuring 365x365 pixels, with minor overlapping sections along both horizontal and vertical axes (refer to Figure 4).

At Batavia weave, the finalized dataset used to train the models comprised 28,693 images. After a thorough review by TS, 8,782 were confirmed as cases, while 19,911 were designated as controls and 4,367 were considered doubtful and eliminated from the data set. The number of patches is not equal to the number of originals multiplied by 12 due to the review carried out by Textil Santanderina. In the case of Sarga, a total of 18,576 Sarga fabric pieces, each measuring 365x365 units, were obtained. Among these, 18,403 were classified as controls and 173 as cases. Notably, this review process identified instances where a specific slice within an original image may constitute a control rather than a case, as the defect was absent in that particular segment. In Sarga dataset, the total number of patches equals the number of originals multiplied by 12. Lastly, pixel normalization was executed across each image, transitioning the pixel values from the [0-255] range to the [0-1] range through simple division by 255. This normalization approach is known to mitigate numerical instability issues during the model training phase.

The following Table 1 presents the processed dataset according to the type of fabric (Batavia and Sarga). It includes information on the total number of 365X365 images (N), the controls, and the cases.

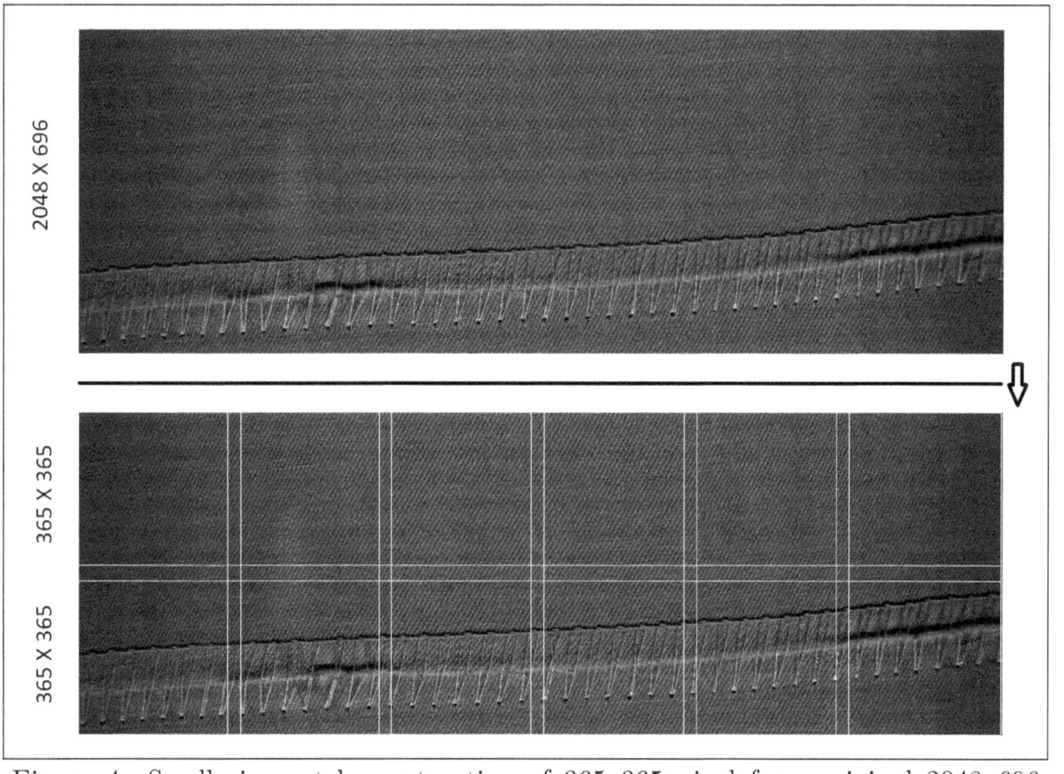

Figure 4: Small size patches extraction of 365x365 pixel from original 2048x696 images, with small overlapping areas in the horizontal and vertical axis.

Fabric	N	Controls	Cases
Batavia	28,693	19,911	8,782
Sarga	18,576	18,403	173

Table 1: The processed dataset includes information on fabrics (Batavia and Sarga), controls and cases.

4.2.2 Data Augmentation

Addressing the class imbalance inherent in the dataset, on-the-fly data augmentation was implemented to augment the number of case samples available for CNN training. This widely employed technique in DL, particularly in computer vision, involves generating modified or transformed variants of existing samples to bolster the training set. By applying vertical and horizontal flips, along with alterations in color and brightness, to the training set, the dataset size was augmented by a factor of five. Consequently, this augmentation strategy augmented the training set with approximately a thousand cases, thereby enriching the diversity and breadth of examples and enhancing the accuracy and generalization capabilities of the models.

4.2.3 Evaluation Strategy

The methodology adopted in this study involves training and evaluating the performance of CNN architectures using a 5-fold stratified cross-validation scheme. In this approach, the dataset is divided into five subsets of equal size, ensuring that the class distribution is maintained across each of them. In each iteration, one fold is reserved as the test set, while the remaining four folds are used for training and validation.

From these four folds, 75% of the data (i.e., 60% of the total dataset) is allocated to training, and the remaining 25% (20% of the total) is used for validation. This 60/20/20 (training/validation/test) split is maintained consistently throughout the five iterations. The validation set is used to monitor the model's performance during training (e.g., for early stopping), but not for the final evaluation. The test set is exclusively used for the final evaluation of performance once training is complete. This process is repeated across the five folds to ensure the robustness of the analysis and mitigate bias in sample selection.

To mitigate the impact of class imbalance during model training, we assigned specific weights to each class based on their relative frequency in the dataset. This was implemented through the class_weight parameter in the model.fit function of Keras, which adjusts the contribution of each class to the binary cross-entropy loss. This approach complements the use of evaluation metrics that are robust to class imbalance, such as the Area Under the ROC Curve (AU-ROC), the Geometric Mean (G-mean), and the Area Under the Precision-Recall Curve (AU-PR).

4.2.4 Metrics

In this study, various complementary performance metrics are analyzed. This approach is adopted due to the highly imbalanced class distribution inherent in the addressed problem. Quality control departments may express particular interest in enhancing the detection capabilities of models for either the negative class (controls or defect-free images), the positive class (cases or images with defects), or both. The assessment of model performance relies on metrics derived from True Positives (TP), True Negatives (TN), False Positives (FP), and False Negatives (FN).

- Accuracy: This metric considers the TP, TN, FP, and FN derived from the predictions generated by a specific method. Its interpretation warrants careful consideration, particularly in the context of imbalanced classification challenges, where it may erroneously convey high performance by favoring the majority class for all input images. The computation of accuracy is delineated by the following 1:

$$Accuracy = (TP + TN)/(TP + TN + FP + FN) \tag{1}$$

- Precision: This metric quantifies the ratio of true positive predictions to all positive predictions generated by a specific method. Essentially, it provides insight into the etrexactness or quality of the method's positive predictions, particularly valuable in scenarios where the ramifications of a false positive are substantial (e.g., medical diagnosis, where a false positive outcome can trigger unnecessary treatment or patient anxiety). The calculation of precision is detailed in the following equation 2:

$$Precision = TP/(TP + FP) \tag{2}$$

- Recall: This metric evaluates the ratio of true positive predictions to all actual positive instances within the dataset. Essentially, it offers insight into the comprehensiveness or quantity of the model's positive predictions, particularly valuable in contexts where the consequences of a false negative are significant (e.g., in fraud detection, where a false negative can lead to financial losses). Its calculation is delineated in the equation 3:

$$Recall = TP/(TP + FN) \tag{3}$$

- F1-score: It represents the harmonic mean between the Precision and Recall metrics. A higher value closer to 1 indicates superior performance of the model. This metric is calculated according to the equation 4:

$$F1 = 2 * (Recall * Precision)/(Recall + Precision) \tag{4}$$

- Geometric Mean: G-mean is a metric used to evaluate the performance of a classification model, particularly in cases of imbalanced datasets. It is designed to balance the accuracy of both the positive and negative classes, ensuring that the model performs well on both sides of the classification task. The calculation is showed in the following equation 5:

$$\text{G-mean} = \sqrt{\frac{TP \times TN}{(TP + FN) \times (TN + FP)}} \tag{5}$$

- Area Under the Precision-Recall Curve (AU-PR): It is a performance metric primarily used in imbalanced classification problems. Unlike accuracy, which can be misleading in the case of imbalanced datasets, AU-PR focuses on evaluating the model's ability to correctly identify positive instances. The Precision-Recall curve plots precision against recall at various threshold levels, and the AU-PR represents the area under this curve. A higher AU-PR indicates a better trade-off between precision and recall, with values closer to 1 indicating superior model performance. This metric is especially useful in cases where the negative class is significantly more frequent than the positive class, such as rare disease detection or fraud detection.

- Area Under the Receiver Operating Characteristic Curve (AU-ROC): It is a widely used metric to evaluate the overall performance of a classification model, especially in binary classification tasks. The ROC curve plots the True Positive Rate (recall) against the False Positive Rate (1-specificity) at various classification thresholds. The area under this curve, known as AUROC, quantifies the model's ability to discriminate between the positive and negative classes. An AUROC value of 1 represents perfect classification, while a value of 0.5 suggests a model with no discriminative power, similar to random guessing. AUROC is particularly useful for comparing models, as it is independent of the classification threshold, making it a robust metric across different decision thresholds.

These metrics were selected for their robustness in imbalanced classification scenarios, which complements the class reweighting strategy adopted during training (as described in Section 4.2.3).

4.2.5 Computational and Implementation Environment

All experiments were conducted on a GPU cluster funded by the InfraRed program of the Junta de Castilla y León. This infrastructure is designed for artificial intelligence and deep learning projects, supporting large-scale experimentation. It included 576 CPU cores (24 processors with 24 cores each) based on the AMD Epyc platform, 9.5 TB of DDR4 RAM, and 320 TB of storage. The cluster was equipped with 30 NVIDIA Tesla A100 GPUs, each with 40 GB of dedicated memory, providing a total computing power of 9360 TFLOPS in FP16 and 4680 TFLOPS in TF32 precision.

A software environment configured on the cluster was used, based on Python 3.9.18, which includes the main libraries required for model training and evaluation. These included TensorFlow, PyYAML, Pandas, NumPy, Scikit-learn, Matplotlib, and Pillow. These tools enable deep learning tasks to be performed using the GPU acceleration provided by the cluster infrastructure.

5 Results

The results obtained by the DL models that have been applied are described in this section, organized according to the weave types, Batavia and Sarga. The pretrained models have been fine-tuned for the specific task of defect detection in these weave types. The metrics presented to evaluate the models are accuracy, precision, recall, F1 score, G-mean, AU-PR and AU-ROC. Additionally, confusion matrices are provided to offer a detailed view of the classification outcomes and to better analyze the behavior of the models with respect to true and false positives and negatives for each class.

Radar plots, also known as spider or star plots, are graphical representations used to visualize multivariate data in a compact and informative manner. In the context of automatic textile inspection, radar plots offer a valuable tool for assessing the performance of various models across multiple metrics. Each axis of a radar plot represents a different metric, such as Accuracy, Precision, Recall, F1-score, AU-PR and AU-ROC allowing for a comprehensive evaluation of model performance. By plotting the values of these metrics on the corresponding axes and connecting them with lines or shapes, radar plots provide a visual summary of how well a model performs across different aspects of the inspection task. In our analysis, radar plots were generated to compare the different performance of pre-trained CNN models against baseline models. These plots allowed us to observe trends in performance metrics and identify areas where certain models excelled or underperformed. Furthermore, radar plots facilitate the comparison of multiple models by presenting their performance side by side. This enables researchers and practitioners to quickly assess the strengths and weaknesses of each model and make informed decisions regarding model selection. Overall, radar plots serve as powerful tools for evaluating the effectiveness of DL models in automatic textile inspection, providing insights that can inform future research and development efforts in this field.

The detailed outcomes for each weave type are provided in the subsequent sections, allowing for a thorough evaluation of the methodologies applied.Also, radar plots have been generated separately for the two types of weave under analysis, Batavia and Sarga.

5.1 Batavia weave

Prior to presenting the evaluation metrics, the confusion matrices corresponding to the Batavia fabric are included to provide a clearer understanding of the models' classification performance. These matrices illustrate the distribution of true positives (TP), true negatives (TN), false positives (FP), and false negatives (FN), offering insight into model behavior under class imbalance conditions.

The confusion matrices for the different models applied to the Batavia fabric are shown in Table 2.

Red	Threshold	TN	FN	TP	FP
VGG16	0.375	8,346	2,129	6,653	11,565
VGG19	0.625	15,290	3,003	5,779	4,621
InceptionV3	0.500	11,430	2,539	6,243	8,481
ResNet101	0.450	15,078	3,649	5,133	4,833
ResNet50V2	0.375	13,927	2,479	6,303	5,984
Xception	0.275	6,798	1,456	7,326	13,113
DenseNet121	0.325	16,399	2,913	5,869	3,512
EfficientNetB0	0.450	15,601	3,848	4,934	4,310
EfficientNetB3	0.325	19,110	4,935	3,847	801

Table 2: Confusion matrices for the different models applied to the Batavia fabric.

Following the implementation of DL models, including the baseline model, on Batavia fiber, the resulting metrics (accuracy, precision, recall, F1-score, G-mean, AU-PR and AU-ROC) detailed in Table 3.

Red	Accuracy	Precision	Recall	F1-score	G-mean	AU-PR	AU-ROC
Baseline	0.70	**1.00**	0.03	0.05	0.16	0.66	0.51
VGG16	0.52	0.37	0.76	0,49	0.56	0.54	0.68
VGG19	0.73	0.56	0.66	0.60	0.71	0.72	0.81
InceptionV3	0.62	0.42	0.71	0.53	0.64	0.61	0.73
ResNet101	0.70	0.52	0.58	0.55	0.67	0.64	0.75
ResNet50V2	0.71	0.51	0.72	0.60	0.71	0.68	0.79
Xception	0.49	0.36	**0.83**	0.50	0.53	0.65	0.76
DenseNet121	0.78	0.63	0.67	**0.65**	**0.74**	**0.80**	**0.87**
EfficientNetB0	0.72	0.53	0.56	0.55	0.66	0.70	0.79
EfficientNetB3	**0.80**	0.83	0.44	0.57	0.65	0.76	0.84

Table 3: Performance metrics of various pre-trained models fine-tuned on the Batavia weave dataset. The table compares the models across multiple metrics, including Accuracy, Precision, Recall, F1-score, G-mean, AU-PR (Area Under the Precision-Recall curve), and AU-ROC (Area Under the Receiver Operating Characteristic curve).

Based on the performance metrics in the Table 3, DenseNet121 stands out as the most effective neural network model for the given task. DenseNet121 model exhibits

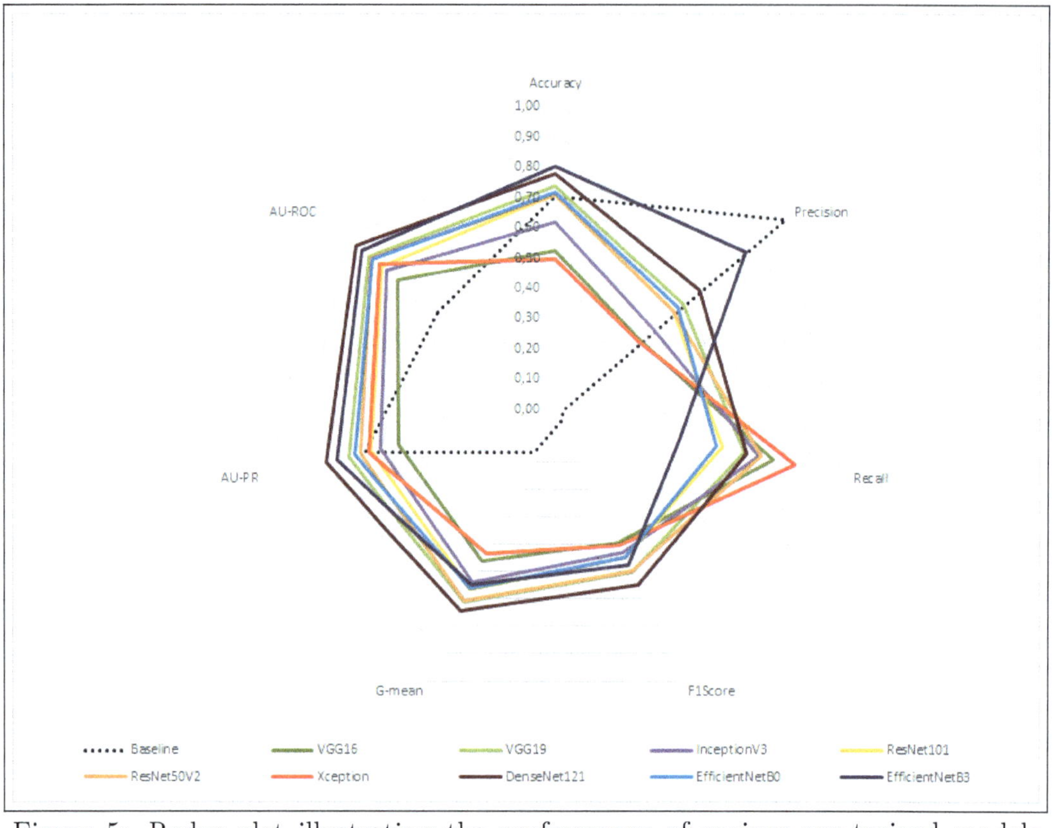

Figure 5: Radar plot illustrating the performance of various pre-trained models fine-tuned on the Batavia weave dataset. The plot compares the models across key metrics, including Accuracy, Precision, Recall, F1-score, G-mean, AU-PR, and AU-ROC.

the best overall performance, excelling across several key metrics. It achieves the highest AU-ROC (0.87), indicating the best ability to discriminate between positive and negative classes. Additionally, it demonstrates a relatively high F1-score of 0.65, indicating a good balance between precision and recall. The model also achieves a precision of 0.63, which is strong and well-balanced. Its recall is 0.67, showing that a significant proportion of positive instances are correctly identified. The AU-PR is 0.80, reflecting strong performance in dealing with imbalanced classes.

In contrast, the Baseline (human) model shows considerably lower performance across most metrics. The precision of the Baseline is perfect at 1.00, but this comes at the cost of a very low recall of 0.03, meaning it detects almost no positive instances. As a result, the F1-score is extremely low at 0.05, as the high precision is insufficient to compensate for the poor recall. The AU-ROC is 0.51, which indicates that the human model performs similarly to random guessing in distinguishing between classes. The AU-PR is 0.66, which is better than the AU-ROC but still reflects suboptimal performance compared to the neural networks.

Other models that perform well include EfficientNetB3, which achieves an excellent AU-ROC of 0.84, although its recall (0.44) and F1-score (0.58) are lower than those of DenseNet121. Additionally, VGG16 and VGG19 both perform well in terms of recall (0.76 for VGG16 and 0.66 for VGG19), but their F1-scores are lower than that of DenseNet121.

This section also presents a detailed analysis of the radar plots generated to evaluate the performance of the models applied to Batavia weave. The radar plot represents different performance metrics, offering a clear and concise visualization of how the models perform in key aspects of textile inspection. The trends observed in the data are examined in detail, including differences between pre-trained models and models trained from scratch, as well as the implications of these differences in terms of accuracy and defect detection capability.

The figure 5 shows a radar plot for each one of the studied metrics and the applied models on Batavia weave. Metrics associated to the baseline have also been included (represented with a dashed line).

5.2 Sarga weave

The confusion matrices for the Sarga fabric are presented first, providing a clear overview of the model's performance in distinguishing between defect and non-defect cases.

Table 4 shows the confusion matrices for the different models applied to the Sarga fabric.

Red	Threshold	TN	FN	TP	FP
VGG16	0.725	18,362	116	57	41
VGG19	0.950	18,187	119	54	216
InceptionV3	0.925	18,360	111	62	43
ResNet101	0.850	18,361	107	66	42
ResNet50V2	0.950	17,886	104	69	517
Xception	0.950	18,192	86	87	211
DenseNet121	0.600	14,510	53	120	3,893
EfficientNetB0	0.775	18,376	152	21	27
EfficientNetB3	0.550	18,244	148	25	159

Table 4: Confusion matrices for the different models applied to the Sarga fabric.

After applying DL models, including the baseline model, to Sarga fiber, the resulting metrics like accuracy, precision, recall, F1-score, G-mnean, AU-PR and AU-ROC are presented in Table 5.

Red	Accuracy	Precision	Recall	F1-score	G-mean	AU-PR	AU-ROC
Baseline	**0.99**	0.00	0.00	0.00	0.16	0.50	0.50
VGG16	**0.99**	0.58	0.33	0.42	0.57	0.42	0.86
VGG19	0.98	0.20	0.31	0.24	0.56	0.42	0.91
InceptionV3	**0.99**	0.59	0.36	0.45	0.60	0.52	**0.93**
ResNet101	**0.99**	**0.61**	0.38	**0.47**	0.62	**0.54**	0.90
ResNet50V2	0.97	0.12	0.40	0.18	0.62	0.39	**0.93**
Xception	0.98	0.29	0.50	0.37	0.71	0.40	0.88
DenseNet121	0.79	0.03	**0.69**	0.06	**0.74**	0.50	0.92
EfficientNetB0	**0.99**	0.44	0.12	0.19	0.35	0.29	0.78
EfficientNetB3	0.98	0.14	0.14	0.14	0.38	0.30	0.84

Table 5: Performance metrics of various pre-trained models fine-tuned on the Sarga weave dataset. The table compares the models across multiple metrics, including Accuracy, Precision, Recall, F1-score, G-mean, AU-PR (Area Under the Precision-Recall curve), and AU-ROC (Area Under the Receiver Operating Characteristic curve).

The analysis of the performance metrics reveals that the InceptionV3 model is the most suitable for the task of defect classification in Sarga weave threads. While ResNet101 achieves a slightly higher F1-score (0.47), InceptionV3 combines a competitive F1-score of 0.45 with the highest AU-ROC (0.93), reflecting superior overall discriminative ability across all thresholds. This model maintains a solid balance between precision (0.59) and recall (0.36), allowing it to identify a significant proportion of positive cases while effectively controlling false positives. In contrast, the human baseline, while showing a high Accuracy of 0.99 due to the dominance of the negative class, fails to detect any positive cases, with Precision and Recall values of 0. This highlights that, despite its seemingly high accuracy, the baseline provides no meaningful value in identifying defects. Compared to other models, such as EfficientNetB3 or DenseNet121, InceptionV3 demonstrates a more consistent performance, overcoming the challenges posed by class imbalance that adversely affect precision and recall in these alternatives.

Below, a comprehensive analysis is conducted of the radar plots generated to evaluate the performance of the models applied to Sarga weave.

The figure 6 shows a radar plot for each one of the studied metrics and the applied models for Sarga weave.

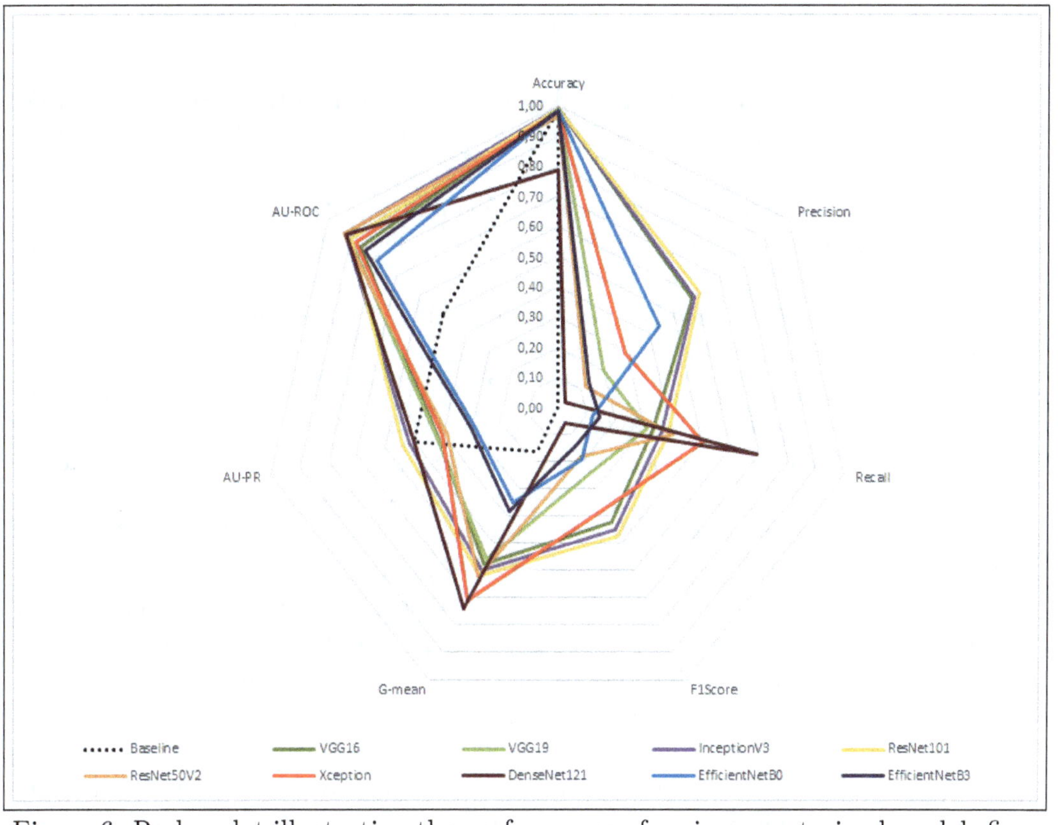

Figure 6: Radar plot illustrating the performance of various pre-trained models fine-tuned on the Sarga weave dataset. The plot compares the models across key metrics, including Accuracy, Precision, Recall, F1-score, G-mean, AU-PR, and AU-ROC.

6 Conclusions

This study has addressed the critical challenge of quality inspection in the textile domain through the application of DL models to oversee fabric quality at TS. The investigation demonstrated that pretrained networks can significantly enhance the inspection process, effectively bridging the gap between the traditional inspection rate of 20 m/min and the production speed of 42 m/min. Our findings revealed that selecting appropriate CNN architectures was pivotal in optimizing defect detection and reducing reliance on human intervention. Consequently, the integration of these AI-based systems has shown great potential to mitigate production interruptions, reduce delays, minimize resource wastage, and manage the high variability of textile products handled by TS. The results of this study underscore the transformative potential of DL in revolutionizing textile quality control, offering a robust and scalable solution to meet the industry's evolving demands.

In conclusion, our study highlights key aspects regarding the effectiveness of DL models, particularly pretrained CNN architectures, for automated textile inspection. First, we emphasize the critical role of selecting appropriate evaluation criteria, as evidenced by the dependency of model performance on the chosen criterion. The key metrics for evaluating the performance of the networks on both fabrics are the F1-score, G-mean, and AU-ROC, as they provide a comprehensive assessment of the models' ability to balance precision and recall (F1-score), maintain a high true positive rate while minimizing false positives and negatives (G-mean), and effectively distinguish between positive and negative classes (AU-ROC). DenseNet121 stands out as the best network for Batavia weave, while InceptionV3 is the most suitable for Sarga weave. These networks exhibit balanced and superior performance across the key metrics, making them optimal for their respective fabrics

Furthermore, our findings highlight the effectiveness of pretrained CNN models in detecting defects across various weave types. By employing radar plots to visualize performance metrics, we provide a comprehensive view of the relative performance of different models. This study significantly contributes to advancing DL techniques in textile quality control, laying the foundation for future research in this field.

Additionally, our study demonstrates the suitability of the proposed DL architecture, specifically CNNs, for online fabric quality control, surpassing the capabilities of human operators. Notably, our analysis reveals superior results compared to the baseline, particularly in Precision and F1-score metrics, highlighting the limitations of traditional metrics such as Accuracy in imbalanced datasets. The integration of DL models into production machinery promises to revolutionize quality control and traceability processes, minimizing human intervention and improving defect detection coverage in textile manufacturing processes. Overall, our study predicts a positive technological impact on textile manufacturing systems, facilitating the transition towards automated quality control and enhanced traceability.

For future work, exploring novel DL architectures specifically designed for textile defect detection and incorporating advanced techniques such as attention mechanisms and reinforcement learning could further improve the performance of automated inspection systems. Additionally, investigating the transferability of models trained on one fabric type to another, as well as the robustness of the models to variations in lighting conditions and fabric textures, would be valuable for real-world applications. Furthermore, integrating real-time feedback mechanisms into inspection systems to enable continuous model refinement and adaptation to evolving manufacturing environments represents an exciting avenue for future research. Currently, the company's detection process is carried out offline on a system with adequate computing power. Nonetheless, we recognize the potential benefits of implementing real-time detection in the production line, and therefore, we plan to explore this possibility in the future. Achieving real-time detection will require overcoming challenges related to preprocessing and inference, ensuring that the system operates efficiently and reliably in a production environment. Another promising direction involves the use of semantic segmentation techniques, which would allow not only for detecting whether a defect exists but also for precisely localizing the defective area within the fabric image. This added spatial information could significantly enhance the interpretability and usability of automated inspection systems in industrial settings.

References

[1] A. Diez-Olivan, J. Del Ser, D. Galar, B. Sierra, Data fusion and machine learning for industrial prognosis: Trends and perspectives towards industry 4.0, Information Fusion 50 (2019) 92–111.

[2] O. Ongbali Samson, A. Afolalu Sunday, Y. Salawu Enesi, Bottleneck problems arising in inter-industry production setting and vertical integration: a review, Technology 10 (5) (2019) 606–612.

[3] S. Shahrabadi, Y. Castilla, M. Guevara, L. G. Magalhães, D. Gonzalez, T. Adão, Defect detection in the textile industry using image-based machine learning methods: a brief review, in: Journal of Physics: Conference Series, no. 1, IOP Publishing, 2022, p. 012010.

[4] A. Saberironaghi, J. Ren, M. El-Gindy, Defect detection methods for industrial products using deep learning techniques: A review, Algorithms 16 (2) (2023).

[5] Z. Jan, F. Ahamed, W. Mayer, N. Patel, G. Grossmann, M. Stumptner, A. Kuusk, Artificial intelligence for industry 4.0: Systematic review of applications, challenges, and opportunities, Expert Systems with Applications 216 (2023) 119456.

[6] M. P. Sikka, A. Sarkar, S. Garg, Artificial intelligence (ai) in textile industry operational modernization, Research Journal of Textile and Apparel 28 (1) (2024) 67–83.

[7] S. Boopathi, U. K. Kanike, Applications of artificial intelligent and machine learning techniques in image processing, in: Handbook of Research on Thrust Technologies' Effect on Image Processing, IGI Global, 2023, pp. 151–173.

[8] M. M. Taye, Understanding of machine learning with deep learning: architectures, workflow, applications and future directions, Computers 12 (5) (2023) 91.

[9] J. J. A. Kovilpillai, S. Jayanthy, An optimized deep learning approach to detect and classify defective tiles in production line for efficient industrial quality control, Neural Computing and Applications (2023) 1–20.

[10] H. Stephani, T. Weibel, R. Rösch, A. Moghiseh, Challenges and approaches when realizing online surface inspection systems with deep learning algorithms, Discover Data

1 (1) (2023) 3.

[11] M. Krichen, Convolutional neural networks: A survey, Computers 12 (8) (2023) 151.

[12] Y. Gao, L. Gao, X. Li, A hierarchical training-convolutional neural network with feature alignment for steel surface defect recognition, Robotics and Computer-Integrated Manufacturing 81 (2023) 102507.

[13] X. Tao, D. Zhang, W. Ma, X. Liu, D. Xu, Automatic metallic surface defect detection and recognition with convolutional neural networks, Applied Sciences 8 (9) (2018).

[14] H. Q. T. Ngo, Design of automated system for online inspection using the convolutional neural network (cnn) technique in the image processing approach, Results in Engineering 19 (2023) 101346.

[15] H. Zhang, W. Xiong, S. Lu, M. Chen, L. Yao, Qa-ustnet: Yarn-dyed fabric defect detection via u-shaped swin transformer network based on quadtree attention, Textile Research Journal 0 (0) (2023) 00405175231158134.

[16] P. R. Jeyaraj, E. R. S. Nadar, Effective textile quality processing and an accurate inspection system using the advanced deep learning technique, Textile research journal 90 (9-10) (2020) 971–980.

[17] B. Wei, B. Xu, K. Hao, L. Gao, Textile defect detection using multilevel and attentional deep learning network (mlma-net), Textile Research Journal 92 (19-20) (2022) 3462–3477.

[18] I. Koulali, M. T. Eskil, Unsupervised textile defect detection using convolutional neural networks, Applied Soft Computing 113 (2021) 107913.

[19] H. Zhang, W. Zhang, Y. Wang, S. Lu, L. Yao, X. Chen, Colour-patterned fabric-defect detection using unsupervised and memorial defect-free features, Coloration Technology 138 (6) (2022) 602–620.

[20] N. Velasco-Pérez, S. Lozano-Juárez, B. Gil-Arroyo, J. M. Sanz, N. Basurto, D. Urda, Á. Herrero, Defect detection in batavia woven fabrics by means of convolutional neural networks, in: P. García Bringas, H. Pérez García, F. J. Martínez de Pisón, F. Martínez Álvarez, A. Troncoso Lora, Á. Herrero, J. L. Calvo Rolle, H. Quintián, E. Corchado (Eds.), 18th International Conference on Soft Computing Models in Industrial and Environmental Applications (SOCO 2023), Springer Nature Switzerland, Cham, 2023, pp. 205–215.

[21] B. Gil-Arroyo, N. Velasco-Pérez, N. Basurto, J. M. Sanz, D. Urda, A. Herrero, Original and processed dataset of batavia and sarga woven fabric images, dataset (2025).
URL `https://doi.org/10.36443/10259/9965`

[22] B. Gil-Arroyo, J. M. Sanz, Ángel Arroyo, D. Urda, N. Basurto, Álvaro Herrero, Dataset for defect detection in textile manufacturing, Data in Brief 59 (2025) 111451. `doi:https://doi.org/10.1016/j.dib.2025.111451`.
URL `https://www.sciencedirect.com/science/article/pii/S2352340925001830`

[23] J. Gu, Z. Wang, J. Kuen, L. Ma, A. Shahroudy, B. Shuai, T. Liu, X. Wang, G. Wang, J. Cai, et al., Recent advances in convolutional neural networks, Pattern recognition 77 (2018) 354–377.

[24] K. Simonyan, Very deep convolutional networks for large-scale image recognition, arXiv preprint arXiv:1409.1556 (2014).

[25] C. Szegedy, V. Vanhoucke, S. Ioffe, J. Shlens, Z. Wojna, Rethinking the inception architecture for computer vision, in: Proceedings of the IEEE conference on computer vision and pattern recognition, 2016, pp. 2818–2826.

[26] K. He, X. Zhang, S. Ren, J. Sun, Deep residual learning for image recognition, in: Proceedings of the IEEE conference on computer vision and pattern recognition, 2016, pp. 770–778.

[27] K. He, X. Zhang, S. Ren, J. Sun, Identity mappings in deep residual networks, in: Computer Vision–ECCV 2016: 14th European Conference, Amsterdam, The Netherlands, October 11–14, 2016, Proceedings, Part IV 14, Springer, 2016, pp. 630–645.

[28] F. Chollet, Xception: Deep learning with depthwise separable convolutions, in: Proceedings of the IEEE conference on computer vision and pattern recognition, 2017, pp. 1251–1258.

[29] G. Huang, Z. Liu, L. Van Der Maaten, K. Q. Weinberger, Densely connected convolutional networks, in: Proceedings of the IEEE conference on computer vision and pattern recognition, 2017, pp. 4700–4708.

[30] M. Tan, Q. Le, Efficientnet: Rethinking model scaling for convolutional neural networks, in: International conference on machine learning, PMLR, 2019, pp. 6105–6114.

 Received 24 January 2025

Type-Theory of Algorithms with Chain-Free Memory and Iterative Computations

Roussanka Loukanova
Institute of Mathematics and Informatics, Bulgarian Academy of Sciences, Bulgaria
rloukanova@gmail.com

Abstract

I introduce an extended version of the Type-Theory of Acyclic Recursion and Acyclic Algorithms (TTAR / TTAA) and its reduction calculi, by adding an extended chain-reduction rule. TTAA provides specialized recursion terms for stepwise, mutually recursive computations. The results of the recursive computations are saved in memory slots, according to assignments, and can be reused, by accessing the corresponding memory slots. The chain-like assignments copy values of terms from one memory slot to another, without any other essential algorithmic steps. The extended reduction calculus eliminates repeated, chain-like assignments of copies of functional values in memory slots. The primary applications of the chain-free type theory of recursion are for algorithmic semantics of formal and natural languages, including of programming languages and parts of compilers that transform recursive programs into iterative ones.

Keywords: formal language, algorithms, lambda, recursion, reduction calculus, computing in memory, memorisation, chain-free computation

1 Background on Moschovakis Theory of Recursion

In a sequence of papers, Yiannis N. Moschovakis introduced a class of formal languages of recursion (FLR), as an approach to the mathematical notion of algorithm. e.g., see [22, 23, 24] on the untyped FLR.

The focus of Moschovakis [25] is introducing a typed version of FLR and type theory L_{ar}^{λ} of algorithms, for potential applications to algorithmic semantics of natural language (NL), covering human language.

The target is a mathematical foundation of algorithms for theoretical and advanced technological applications, by distinguishing between denotational values of language expressions and their algorithmic computations, in similar ways as in semantics of programing languages.

For detailed, mathematical introduction to the original L_{ar}^{λ} and developments of extended type theory L_{ar}^{λ} having correspondingly extended reduction calculi, and more subtle applications of algorithmic semantics, see, e.g., Loukanova [6, 7, 11, 12, 10, 13, 14, 16, 17, 18, 19]. The main focus of the work in these papers, including the present one, is on theoretical developments of the type theory of recursion and algorithms (TTRA) via its own algorithmic syntax-semantics interface. TTRA has innate, algorithmic syntax-semantics interface at its object level via its formal languages and reduction calculi.

In addition, externally, the varieties of TTRA have direct applications to computational, i.e., algorithmic syntax-semantics interfaces of formal and natural languages. The applications of TTRA to formal languages include mathematical foundation of some of the components of contemporary, automatically generated compilers of programming languages. In practical aspects, TTRA provides reduction calculi for underlying parts of compilers that convert recursive programs into more efficient, iterative ones. More generally, TTRA provides algorithmic semantics of programming and specification languages. This sets the perspective of applications of TTRA to automatic verification and proof systems.

Realistic applications of TTRA is to natural language processing (NLP) and algorithmic syntax-semantics interface in computational grammar of natural language. This is demonstrated, e.g., by Loukanova [6, 7, 12, 10, 17, 18, 19].

Loukanova [13] demonstrates a perspective for applications to computational neuroscience.

Among the existing applications of type-theory of acyclic algorithms L_{ar}^{λ} is for algorithmic syntax-semantics in large-scale computational grammars of natural language. In particular, Head-Driven Phrase Structure Grammar (HPSG) approach to formal grammar of natural language has been including semantic representations in its large scale grammars, by using a technique called Minimal Recursion Semantics (MRS), e.g., see Copestake et al. [1]. Type theory of acyclic recursion L_{ar}^{λ} provides mathematical foundation of MRS, via currying encoding of relations between objects.

The various areas of Artificial Intelligence (AI) have needs of algorithmic syntax-semantics interfaces, e.g., in extraction and use of semantic information. Type theory of acyclic and full recursion L_{ar}^{λ} / L_{r}^{λ} has potential in these areas as well.

The type theory of recursion and algorithms (TTRA, TTR, TTA) L_{r}^{λ} and its sub-varieties, e.g., type theory of acyclic recursion and acyclic algorithms (TTARA)

L_{ar}^{λ} have potentials for applications to intelligent computational systems of information, especially for integration in AI software systems, in robotics, that perform algorithmic procedures. TTA provides computational foundations of AI, and targets to provide AI with efficient and reliable performance.

The varieties and extensions of type theory L_{ar}^{λ} provide mathematical and algorithmic foundations of, and especially useful applications to computational syntax-semantics of natural and formal languages. Such applications include programming and other specification languages in Computer Science and AI. In the above and forthcoming sequences of papers, I have already established applications of type theory of recursion and algorithms to Natural Language Processing (NLP) and computational grammars that cover computational syntax-semantics interfaces. The above papers cover type theory TTA and TTAA in functional approaches, via currying coding.

For relational versions of Dependent-Type Theory of Situated Information (DTTSI), see, e.g., [3, 4, 5, 9]. For an integration of the DTTSI with numerical and statistical data, see [15], which introduces a potent hybrid type theory for applications to data science and computational grammar.

The relational DTTSI directly provides mathematical and computational foundations of semantics and syntax-semantics interfaces in varieties of approaches to formal and computational grammar approaches. These include approaches such as presented in [28]. Importantly, DTTSI and its functional variaties L_{ar}^{λ} provide mathematical formalization of the semantic representation by Minimal Recursion Semantics (MRS) in Head-Driven Phrase Structure Grammar (HPSG), see [1] and [26].

Moschovakis Type Theory of Acyclic Recursion (TTAR) L_{ar}^{λ} was introduced in [25], via its demonstration with examples for possible applications to computational semantics of natural language. Moschovakis [25] provides a mathematical proof that the formal theory L_{ar}^{λ} is a proper extension of TY_2 introduced by Gallin [2], and thus, of Montague's Intensional Logic (IL) [21]. Furthermore, Loukanova [17] extends that proof, by mathematical proofs and by applications to syntax-semantics representations of natural language. The Gallin type system of L_{ar}^{λ} in [25], especially polymorphic types [6], provides a computational facility for formalisation of states, e.g., of computer memory, and other context information.

Removing the Acyclicity Constraint (AC) on the formal language of L_{ar}^{λ}, see later Def. 1 of the L_{ar}^{λ} terms, yields the type theory of full recursion (TTFR) L_{r}^{λ}. The type theory of full recursion (TTFR) provides type-theoretic algorithmic semantics of natural and formal languages by generalizing the let-expressions of lambda-calculus, in particular, of the formal language LCF, by Scott [29] and Plotkin [27]. In [17], I provide a mathematical proof that Scott let-terms are not algorithmically equivalent with the where-recursion terms of the TTFR L^{λ}_{r}. They remain denotationally equivalent.

This paper provides a substantial and important extension of the chain rule introduced in [20]. In Sect. 2, I provide a brief introduction to Moschovakis Type-Theory of Acyclic Recursion, L^{λ}_{ar}, which is more in line with its recent developments and new directions of its applications, see [6, 7, 11, 12, 13, 16, 17, 18, 19].

Section 3 provides the reduction rules of L^{λ}_{ar}, introduced by [25]. I provide them also in abbreviated forms by vector notations, for sequences of variables, abstractions, and assignments. In the second part of the paper, Sect. 5, I extend the reduction calculus of L^{λ}_{ar}, for reducing assignment chains that copy information from one memory location (slot) to another, without any algorithmically significant changes in it. Also, I provide some theoretical results. In Sects. 4–5, I demonstrate the role of the canonical forms of the L^{λ}_{ar} terms for their algorithmic and denotational semantics by examples of *renderings*, i.e., translations, of natural language (NL) expressions into the extended L^{λ}_{ar}, with and without chain assignments.

2 Introduction to Moschovakis Type-Theory of Acyclic Recursion L^{λ}_{ar}

2.1 Syntax

The set $\mathsf{Types}_{L^{\lambda}_{ar}}$ is defined recursively by the rules (in Backus-Naur Form):

$$\tau ::= \mathsf{e} \mid \mathsf{t} \mid \mathsf{s} \mid (\tau \to \tau) \qquad \text{(Types)}$$

The types e, t, s are for entities, truth values, and states, respectively. The abbreviations in (1a)–(1c) are, correspondingly, for state-dependent types $\widetilde{\sigma}$, $\widetilde{\mathsf{t}}$, $\widetilde{\mathsf{e}}$. The type (1d) is for n-place functions, by currying coding:

$$\widetilde{\sigma} \equiv (\mathsf{s} \to \sigma) \quad \text{(for state-dependent types } \sigma\text{)}, \tag{1a}$$

$$\widetilde{\mathsf{e}} \equiv (\mathsf{s} \to \mathsf{e}) \quad \text{(for state-dependent entities)}, \tag{1b}$$

$$\widetilde{\mathsf{t}} \equiv (\mathsf{s} \to \mathsf{t}) \quad \text{(for state-dependent entities)}, \tag{1c}$$

$$(\overrightarrow{\tau} \to \sigma) \equiv (\tau_1 \to \cdots \to (\tau_n \to \sigma)) \in \mathsf{Types} \tag{1d}$$

Typed Vocabulary of L^{λ}_{ar} The formal language of L^{λ}_{ar} has, for every $\tau \in \mathsf{Types}$:

(1) denumerable sets of typed *constants*, $K_\tau = \mathsf{Consts}_\tau = \{\mathsf{c}^{\tau}_0, \ldots, \mathsf{c}^{\tau}_k, \ldots\}$

(2) two kinds of infinite, denumerable sets of typed variables:

pure variables: $\mathsf{PureVars}_\tau = \{\mathsf{v}_0^\tau, \mathsf{v}_1^\tau, \dots\}$

recursion (memory) variables: $\mathsf{RecVars}_\tau = \{\mathsf{r}_0^\tau, \mathsf{r}_1^\tau, \dots\}$

I call the recursion variables also *memory variables* or *memory locations (slots, cells)*, for their role in algorithmic computations. Values, which are obtained by recursive, or iterative, computations, can be saved, i.e., memorised, in typed memory locations.

Definition 1 (Terms). *For each* $\tau \in \mathsf{Types}$, Terms_τ *is the set of the terms of type* τ. $\mathsf{Terms} = \mathsf{Terms}(\mathrm{L}_{\mathrm{ar}}^\lambda) = \cup\, \mathsf{Terms}_\tau$ *is the set of the terms of* $\mathrm{L}_{\mathrm{ar}}^\lambda$.

$$
\begin{aligned}
A :\equiv\ & \mathsf{c}^\tau : \tau \mid x^\tau : \tau && (\textit{constants and variables}) && \text{(2a)}\\
& \mid B^{(\sigma\to\tau)}(C^\sigma) : \tau && (\textit{application terms}) && \text{(2b)}\\
& \mid \lambda(v^\sigma)(B^\tau) : (\sigma \to \tau) && (\lambda\textit{-abstraction terms}) && \text{(2c)}\\
& \mid A_0^{\sigma_0}\ \mathsf{where}\ \{\, p_1^{\sigma_1} := A_1^{\sigma_1}, \dots, p_n^{\sigma_n} := A_n^{\sigma_n} \,\} : \sigma_0 && (\textit{recursion terms}) && \text{(2d)}
\end{aligned}
$$

given that

(1) $\mathsf{c} \in K_\tau$, $x \in \mathsf{Vars}_\tau = \mathsf{PureVars}_\tau \cup \mathsf{RecVars}_\tau$, $v \in \mathsf{PureVars}_\sigma$

(2) the term (2d) *satisfies the Acyclicity Constraint (AC), and:*

- *(a) for* $n \geq 0$, $i = 0, \dots, n$, $\sigma_i \in \mathsf{Types}$, $A_i \in \mathsf{Terms}_{\sigma_i}$
- *(b) for* $i = 1, \dots, n$, $p_i \in \mathsf{RecVars}_{\sigma_i}$ *are pairwise different recursion variables*

Acyclicity Constraint (AC) The sequence of assignments (3a) is *acyclic* iff there is a function $\mathsf{rank} : \{p_1, \dots, p_n\} \to \mathbb{N}$, such that:

$$
\begin{aligned}
& \{\, p_1^{\sigma_1} := A_1^{\sigma_1}, \dots, p_n^{\sigma_n} := A_n^{\sigma_n} \,\} \quad (n \geq 0) && \text{(3a)}\\
& \text{if } p_j \text{ occurs freely in } A_i, \text{ then } \mathsf{rank}(p_i) > \mathsf{rank}(p_j) && \text{(3b)}
\end{aligned}
$$

Note 1. *Without the Acyclicity Constraint (AC), the rules* (2a)–(2d) *define the language* L_r^λ *of full recursion.*

Definition 2 (Explicit, λ-calculi, Recursion Terms). *For every* $A \in \mathsf{Terms}$*:*

(1) A *is* explicit *iff the constant* where *does not occur in it*

(2) A *is a* λ*-calculus term iff it is explicit and no recursion variable occurs in it*

(3) We call $A \in \mathsf{Terms}$ *of the form* $A \equiv \lambda(\overrightarrow{v_j})(B)$, *with a non-empty* λ*-prefix, i.e.,* $j \geq 1$, *a* λ-abstract *or* λ-term

(4) We call $A \in \mathsf{Terms}$ *of the form* (2d), $n \geq 1$, *a* recursion term *or* where-term.

About the Denotational Semantics of L^{λ}_{ar} The denotation function $\mathsf{den}^{\mathfrak{A}}$ is defined by structural induction on term formation by Def. 1, see [25] and [6, 7, 11, 12, 13, 16, 17, 18, 19].

Definition 3 (Immediate and Pure Terms). *The set of the* immediate terms *consists of all terms of the form* (4), *for* $p \in \mathsf{RecVars}$, $u_i, v_j, \in \mathsf{PureVars}$ *(*$i = 1, \ldots, n$, $j = 1, \ldots, m$*)*, $V \in \mathsf{Vars}$*:*

$$T :\equiv V \mid p(v_1)\ldots(v_m) \mid \lambda(u_1)\ldots\lambda(u_n)p(v_1)\ldots(v_m), \quad \textit{for } m, n \geq 0 \tag{4}$$

Every term A that is non-immediate is pure.

Informally, A term (4), i.e., a variable V or of the form $\lambda(\overrightarrow{u})p(\overrightarrow{v})$ $(m, n \geq 0)$, is immediate if its denotation is obtained by the values provided via the variable valuations $g \in G$, without any algorithmic steps.

3 Reduction Calculus of L^{λ}_{ar}

In this section, I present the reduction rules of L^{λ}_{ar}, originally introduced by Moschovakis [25]. Loukanova [12, 17] extends L^{λ}_{ar} by adding the γ^* rule, Def. 5–6. The reduction rules determine the reduction relation $\Rightarrow$ of L^{λ}_{ar}, which extends the basic relation of congruence $\equiv_{\mathsf{c}}$ in L^{λ}_{ar}. For a full, formal definition, see [17].

Definition 4. *The Congruence Relation is the smallest relation* $\equiv_{\mathsf{c}}$ *between terms that is closed under:*

(1) reflexivity, symmetricity, transitivity, i.e., $\equiv_{\mathsf{c}}$ *is an equivalence relation*

(2) operators of term-formation, by Def. 1

(3) renaming bound pure and recursion variables without variable collisions, i.e., by free replacements

(4) re-ordering of the assignments within the recursion terms

Congruence The congruence relation is maintained by the reduction relation:

$$\text{If } A \equiv_c B, \text{ then } A \Rightarrow B \tag{cong}$$

Transitivity

$$\text{If } A \Rightarrow B \text{ and } B \Rightarrow C, \text{ then } A \Rightarrow C \tag{trans}$$

Compositionality Replacement of sub-terms with correspondingly reduced ones respects the term structure by the definition of the term syntax.

If $A \Rightarrow A'$ and $B \Rightarrow B'$, then $A(B) \Rightarrow A'(B')$ (comp-ap / rep1)

If $A \Rightarrow B$, then $\lambda(u)(A) \Rightarrow \lambda(u)(B)$ (comp-lam / rep2)

If $A_i \Rightarrow B_i$, for $i = 0, \ldots, n$, then (comp-rec / rep3)

$A_0 \text{ where } \{ p_1 := A_1, \ldots, p_n := A_n \} \Rightarrow B_0 \text{ where } \{ p_1 := B_1, \ldots, p_n := B_n \}$

Head Rule If $p_i \neq q_j$ and p_i does not occur freely in B_j, then

$$\begin{aligned} &(A_0 \text{ where } \{ \overrightarrow{p} := \overrightarrow{A} \}) \text{ where } \{ \overrightarrow{q} := \overrightarrow{B} \} \\ &\Rightarrow A_0 \text{ where } \{ \overrightarrow{p} := \overrightarrow{A},\ \overrightarrow{q} := \overrightarrow{B} \} \end{aligned} \tag{head}$$

Bekič-Scott Rule If $p_i \neq q_j$ and q_j does not occur freely in A_i, then

$$\begin{aligned} &A_0 \text{ where } \{ p := (B_0 \text{ where } \{ \overrightarrow{q} := \overrightarrow{B} \}), \overrightarrow{p} := \overrightarrow{A} \} \\ &\Rightarrow A_0 \text{ where } \{ p := B_0, \overrightarrow{q} := \overrightarrow{B},\ \overrightarrow{p} := \overrightarrow{A} \} \end{aligned} \tag{B-S}$$

Recursion-Application Rule If p_i does not occur freely in B, then

$$(A_0 \text{ where } \{ \overrightarrow{p} := \overrightarrow{A} \})(B) \Rightarrow A_0(B) \text{ where } \{ \overrightarrow{p} := \overrightarrow{A} \} \tag{recap}$$

Application Rule If B is a proper term and b is a fresh recursion (memory) variable, then

$$A(B) \Rightarrow\ A(b) \text{ where } \{ b := B \} \tag{ap}$$

λ-Rule If, for every $i = 1, \ldots, n$, p'_i is a fresh recursion (memory) variable, and A'_i is the result of the replacement of all the free occurrences of $p_1, \ldots, p_n$ in A_i with $p'_1(u), \ldots, p'_n(u)$, respectively, i.e.:

$$A'_i \equiv A_i\{p_1 :\equiv p'_1(u), \ldots, p_n :\equiv p'_n(u)\} \equiv A_i\{\overrightarrow{p} :\equiv \overrightarrow{p'(u)}\} \tag{6a}$$

then

$$\begin{aligned} &\lambda(u)\,(A_0 \text{ where } \{ p_1 := A_1, \ldots, p_n := A_n \}) &&(\lambda) \\ &\Rightarrow \lambda(u)\, A'_0 \text{ where } \{ p'_1 := \lambda(u)\, A'_1, \ldots, p'_n := \lambda(u)\, A'_n \} \\ &\equiv \lambda(u) A_0\{\overrightarrow{p} :\equiv \overrightarrow{p'(u)}\} \text{ where } \{ \overrightarrow{p'} := \overrightarrow{\lambda(u) A\{\overrightarrow{p} :\equiv \overrightarrow{p'(u)}\}} \} &&\text{(abbrev)} \end{aligned}$$

γ^*-Rule

$$A \equiv_{\mathsf{c}} A_0 \text{ where } \{ \overrightarrow{a} := \overrightarrow{A}, p := \lambda(\overrightarrow{u})\lambda(v)P, \overrightarrow{b} := \overrightarrow{B} \} \tag{7a}$$

$$\Rightarrow_{\gamma^*} A_0' \text{ where } \{ \overrightarrow{a} := \overrightarrow{A}', p' := \lambda(\overrightarrow{u})P, \overrightarrow{b} := \overrightarrow{B}' \} \tag{γ^*}$$

$$\begin{aligned} \equiv A_0\{ p(\overrightarrow{u})(v) :\equiv p'(\overrightarrow{u}) \} \text{ where } \{ & \overrightarrow{a} := \overrightarrow{A}\{ p(\overrightarrow{u})(v) :\equiv p'(\overrightarrow{u}) \}, \\ & p' := \lambda(\overrightarrow{u})P, \\ & \overrightarrow{b} := \overrightarrow{B}\{ p(\overrightarrow{u})(v) :\equiv p'(\overrightarrow{u}) \} \} \end{aligned} \tag{7b}$$

where

- the term $A \in \mathsf{Terms}$ satisfies the (strict) γ^*-condition (given in Def. 5) for the assignment $p := \lambda(\overrightarrow{u})\lambda(v)P : (\overrightarrow{\vartheta} \to (\vartheta \to \tau))$
- $p' \in \mathsf{RecV}_{(\overrightarrow{\vartheta} \to \tau)}$ is a fresh recursion variable
- for each part X_i of $\overrightarrow{X}$ in (γ^*) and (7b), X_i' is the result of the replacements $X_i' \equiv X_i\{ p(\overrightarrow{u})(v) :\equiv p'(\overrightarrow{u}) \}$ of all occurrences of $p(\overrightarrow{u})(v)$ by $p'(\overrightarrow{u})$, modulo renaming the variables $\overrightarrow{u}, v$, for $i \in \{0, \ldots, n_X\}$

Definition 5 (γ^*-Condition). *See [12]. A recursion term $A \in \mathsf{Terms}$ satisfies the γ^*-condition for an assignment $p := \lambda(\overrightarrow{u}^{\overrightarrow{\vartheta}})\lambda(v^{\vartheta})(P^{\tau}) : (\overrightarrow{\vartheta} \to (\vartheta \to \tau))$, with respect to $\lambda(v)$, if and only if A is of the form:* (8)

$$A \equiv A_0 \text{ where } \{ \overrightarrow{a} := \overrightarrow{A}, \; p := \lambda(\overrightarrow{u})\lambda(v)P, \; \overrightarrow{b} := \overrightarrow{B} \} \tag{8}$$

with the sub-terms of appropriate types, such that the following holds:

(1) $P \in \mathsf{Terms}_{\tau}$ does not have any free occurrences of v in it: $v \notin \mathsf{FreeVars}(P)$

(2) All occurrences of p in A_0, $\overrightarrow{A}$, and $\overrightarrow{B}$ are in sub-terms $p(\overrightarrow{u})(v)$, which are in the scope of $\lambda(\overrightarrow{u})\lambda(v)$, modulo renaming the bound variables $\overrightarrow{u}, v$

Definition 6 (γ^*-Reduction). *The* γ^*-reduction *relation $\Rightarrow^*_{\gamma^*}$ is the smallest relation $\Rightarrow^*_{\gamma^*} \subseteq \mathsf{Terms} \times \mathsf{Terms}$ between terms, which is closed under the γ^*-reduction rules, i.e.,* (cong)–(λ), *and* (γ^*) *rules. Often, when the context makes it clear, we write $A \Rightarrow_{\gamma^*} B$, or even $A \Rightarrow B$ for $A \Rightarrow^*_{\gamma^*} B$. We write $A \Rightarrow_{(\gamma^*)} B$ to highlight that (γ^*)-rule is applied.*

4 Examples for Reductions to Canonical Forms

Example 1. *Assume the renderings in* (9a)–(9b). *Thus, the denotation of the VP* (9b) *is designated by* $\mathsf{den}(B)$. *In* $\mathrm{L}^{\lambda}_{\mathrm{ar}}$ *and its reduction calculi the algorithm for computing* $\mathsf{den}(B)$, *and thus, of the VP* (9b), *is determined by the canonical form* (9g), *which is congruent to* (9h).

$$\textit{[larger than]}_{\mathrm{ADJ}} \xrightarrow{\mathsf{render}} \mathit{isLargerThan} \in \mathsf{Consts}_{(\widetilde{\mathsf{e}}\to(\widetilde{\mathsf{e}}\to\widetilde{\mathsf{t}}))} \tag{9a}$$

$$\textit{[is [larger than [some cube]}_{\mathrm{NP}}\textit{]}_{\mathrm{ADJP}}\textit{]}_{\mathrm{VP}} \xrightarrow{\mathsf{render}} B \in \mathsf{Terms}_{(\widetilde{\mathsf{e}}\to\widetilde{\mathsf{t}})} \tag{9b}$$

$$B \equiv \lambda(y)\Big[[\mathit{some}(\mathit{cube})](\lambda(x_o)\,[\mathit{isLargerThan}(x_o)(y)])\Big] \tag{9c}$$

$$\Rightarrow \lambda(y)\Big[[\mathit{some}(o) \text{ where } \{\, o := \mathit{cube} \,\}](\lambda(x_o)\,[\mathit{isLargerThan}(x_o)(y)])\Big] \tag{9d}$$

by (ap), (comp-ap / rep1), (comp-lam / rep2) *from* (9c)

$$\Rightarrow \lambda(y)[\mathit{some}(o)(\lambda(x_o)\,[\mathit{isLargerThan}(x_o)(y)]) \text{ where } \{\, o := \mathit{cube} \,\}] \tag{9e}$$

by (recap), (comp-lam / rep2) *from* (9d)

$$\Rightarrow \lambda(y)[\mathit{some}(o)(h') \text{ where } \{\, o := \mathit{cube}, \; h' := \lambda(x_o)\,[\mathit{isLargerThan}(x_o)(y)] \,\}] \tag{9f}$$

by (ap), (head), (cong), (comp-lam / rep2) *from* (9e)

$$\Rightarrow_{\mathsf{cf}} \mathsf{cf}(B) \equiv \lambda(y)\mathit{some}(o'(y))(h(y)) \text{ where } \{\, o' := \lambda(y)\mathit{cube}, \;\; h := \lambda(y)\lambda(x_o)\,[\mathit{isLargerThan}(x_o)(y)] \,\} \tag{9g}$$

by (λ) *from* (9f)

$$\equiv_c \lambda(y_k)\mathit{some}(o'(y_k))(h(y_k)) \text{ where } \{\, o' := \lambda(y_k)\mathit{cube}, \;\; h := \lambda(y_k)\lambda(x_o)\mathit{isLargerThan}(x_o)(y_k) \,\} \tag{9h}$$

by renaming congruence (cong) *from* (9g)

$$\Rightarrow_{\gamma^*} \mathsf{cf}_{\gamma^*}(B) \equiv \lambda(y)\mathit{some}(o)(h(y)) \text{ where } \{\, o := \mathit{cube}, \;\; h := \lambda(y)\lambda(x_o)\,[\mathit{isLargerThan}(x_o)(y)] \,\} \tag{9i}$$

from (9g) *by* γ^* *[17]*

Note that, after application of the γ^* *rule [12, 17] to* (9g) *of the original reduction calculus of* $\mathrm{L}^{\lambda}_{\mathrm{ar}}$, *the result is* (9i), *a* γ^**-canonical form, which is more efficient, without unnecessary* λ*-abstractions and applications.*

Example 2. *Let* $A \in \mathsf{Terms}$ *render the sentence in* (10a). *The canonical form* $\mathsf{cf}(A)$ *of* A *determines the algorithm for computing the denotation* $\mathsf{den}(A)$ *of* A *and the* *sentence in* (10a)*:*

$$[\ [N]_{NP}\ [is\ [larger\ than\ [some\ cube]_{NP}]_{ADJP}]_{VP}\ ]_S \xrightarrow{\mathsf{render}} A \in \mathsf{Terms}_{\tilde{t}} \tag{10a}$$

$$A \equiv [\lambda(y)([some(cube)](\lambda(x_o)\,[isLargerThan(x_o)(y)]))](N) \tag{10b}$$

$$\begin{aligned}
&\Rightarrow_{\mathsf{cf}}\ \mathsf{cf}(A) \equiv \underbrace{[\lambda(y_k)(some(o'(y_k))(h(y_k)))](k)}_{explicit,\ irreducible}\ \text{where}\\
&\quad\{\, h := \lambda(y_k)\lambda(x_o)\,[isLargerThan(x_o)(y_k)],\\
&\quad\ o' := \lambda(y_k)cube,\ \ k := N\,\}\\
&\quad \textit{from } (9h),\ \textit{by } (\text{recap}), (\text{ap}), (\text{comp-rec / rep3}), (\text{head}), (\text{cong})
\end{aligned} \tag{10c}$$

$$\begin{aligned}
&\Rightarrow_{\gamma^*}\ \mathsf{cf}_{\gamma^*}(A) \equiv \underbrace{[\lambda(y_k)some(o)(h(y_k))](k)}_{explicit,\ irreducible}\ \text{where}\\
&\quad\{\, h := \lambda(y_k)\lambda(x_o)[isLargerThan(x_o)(y_k)],\ o := cube,\ k := N\,\}\\
&\quad \textit{from } (10c)\ \textit{by } \gamma^*\ [17]
\end{aligned} \tag{10d}$$

The γ^**-canonical form* $\mathsf{cf}_{\gamma^*}(A)$ (10d) *is more efficient algorithmically than* (10c).

5 Reduction Calculus of Chain Free Memory

In this section, I extend the set of the reduction rules of $\mathrm{L}_{\mathrm{ar}}^{\lambda}$ to an extended reduction calculus of $\mathrm{L}_{\mathrm{ar}}^{\lambda}$, by (chain) rule, for removing chain assignments. The new (chain) rule reduces repeated saving, via chain-like term assignments, $q := p$, $p := A$, of the value of a term A, which has already been computed and saved in a primary memory location p. More generally, the (chain) rule removes repeated memorization $q := \lambda(\overrightarrow{y})(p(\overrightarrow{y}))$ of the same, already computed, value of A saved in p, via a chain mode, $q := \lambda(\overrightarrow{y})(p(\overrightarrow{y}))$, $p := A$, modulo λ-abstraction.

At first, I shall motivate the need of the extended reduction calculus with examples.

Example 3. *Assume that* $m, p, i, f_0, m_0 \in \mathsf{RecVars}$. *The tentative denotation values of* $f_0, m_0 \in \mathsf{FreeVars}(F^1)$ *would be provided by variable valuations. In the term* (11a), *the assignments* $p := i$ *and* $i := f_0$ *save and access the tentative value of* f_0 *in chain-like fashion, without contribution of essential information. We would like to reduce this chain of assignments, by the* (chain) *rule that I introduce in this section. Note that after the application of the* (chain) *rule, we still have an underspecified term* F, *by* $f_0, m_0 \in \mathsf{FreeVars}(F)$.

$$F^1 \equiv m(p) \text{ where } \{ p := i,\ i := f_0,\ m := m_0 \} \tag{11a}$$

$$\Rightarrow_{\mathsf{ch}} F \equiv m(i) \text{ where } \{ i := f_0,\ m := m_0 \} \quad \textit{by (chain) rule} \tag{11b}$$

Assume now that $cube, large \in \mathsf{Consts}$.

$$F_{spec} \equiv m(i) \text{ where } \{ i := f_0,\ m := m_0,\ f_0 := cube,\ m_0 := large \} \tag{12a}$$

$$\Rightarrow_{\mathsf{ch}} F'_{spec} \equiv m(f_0) \text{ where } \{ m := m_0,\ f_0 := cube,\ m_0 := large \} \tag{12b}$$

$$\Rightarrow_{\mathsf{ch}} F''_{spec} \equiv m_0(f_0) \text{ where } \{ f_0 := cube,\ m_0 := large \} \quad \textit{by (chain) rule} \tag{12c}$$

Example 4 (A Less Efficient Option). *Let us consider a sentence like "The cube is large" consisting of a subject noun phrase (NP), e.g., a referential definite description "the cube", and a predicative verbal phrase (VP), e.g., "is large". At first, we consider rendering of the components and of the entire sentence into terms, which are not necessarily in canonical forms. The canonical sub-forms, e.g.,* (14b) *of* (14a), *and the final canonical form of the rendering term* E *in* (15a) *can be obtained afterwards, by using the reduction rules.*

$$C_i \equiv the(cube) \Rightarrow_{\mathsf{cf}} the(c) \text{ where } \{ c := cube \} \qquad \textit{by (ap)} \tag{13}$$

The NP "the cube" can be rendered to the term C *or* $\mathsf{cf}(C)$ *in* (13), *or alternatively to* (14a):

$$[the\ cube]_{NP} \xrightarrow{\mathsf{render}} E^0 \equiv_{\mathsf{c}} i \text{ where } \{i := the(cube)\} \tag{14a}$$

$$\Rightarrow i \text{ where } \{i := the(c),\ c := cube \} \quad \textit{by (comp-rec / rep3) from (13), (14a); and (B-S)} \tag{14b}$$

Then $T_{isLarge}(E^0)$ *can be reduced to a canonical form, e.g., as in* (15a)–(15h):

$$[\,[The\ cube]_{NP}, [is\ large]_{VP}\,]_S \xrightarrow{\mathsf{render}} T_{isLarge}(E^0) \equiv_{\mathsf{c}} E \tag{15a}$$

$$E \equiv [b \text{ where } \{ b := isLarge \}](i \text{ where } \{i := the(cube)\}) \tag{15b}$$

$$\Rightarrow [b(i \text{ where } \{i := the(cube)\})] \text{ where } \{ b := isLarge \} \quad \textit{by (recap)} \tag{15c}$$

$$\Rightarrow [b(e) \text{ where } \{ e := [i \text{ where } \{i := the(cube)\}]\, \}] \text{ where } \{ b := isLarge \} \tag{15d}$$

by (ap) *to the head subterm of* (15c); (comp-rec / rep3) *to* (15c)

$$\Rightarrow b(e) \text{ where } \{ e := [i \text{ where } \{i := the(cube)\}], \\ b := isLarge \} \tag{15e}$$
by (head) *to* (15d)

$$\Rightarrow b(e) \text{ where } \{ e := i, \ i := the(cube), \ b := isLarge \} \quad by \text{ (B-S)} \tag{15f}$$

$$\Rightarrow_{\mathsf{ch}} T \equiv b(i) \text{ where } \{ i := the(cube), \ b := isLarge \} \tag{15g}$$
by (chain) *rule from* (15f)

$$\Rightarrow_{\mathsf{cf}} \mathsf{cf}(T) \equiv \mathsf{cf}_{\mathsf{ch}}(T) \equiv b(i) \text{ where } \{ i := the(c), \ c := cube, \\ b := isLarge \} \tag{15h}$$
by (comp-rec / rep3), (B-S), *from* (13), (15g)

To reduce chain-like assignments of recursion terms, as in (15f), I add the new reduction rule (chain) to the set of the reduction rules of $\mathrm{L}^{\lambda}_{\mathrm{ar}}$. The term (15g) doesn't have the chain assignments $e := i$, $i := the(cube)$, after applying the (chain) rule to (15f). As a result, the canonical form (15h) is simplified. Thus, the (chain) rule optimises the canonical forms of the semantic representations of natural language, by providing simpler algorithmic steps.

Chain Rule Assume that $Q, A, A_i \in \mathsf{Terms}$, $q, p, p_j \in \mathsf{RecVars}$, are such that for all $i \in \{0, 1, \ldots, n\}$, $j \in \{1, \ldots, n\}$ $(n \geq 0)$:

(1) Q, A, p, q are of the same type, i.e., for some $\tau \in \mathsf{Types}$, $Q, A \in \mathsf{Terms}_\tau$, $p, q \in \mathsf{RecVars}_\tau$
(2) $Q \in \mathsf{Terms}$, Q is explicit, irreducible
(3) $\mathsf{FreeVars}(Q) = \{ p \}$
(4) $\mathsf{Consts}(Q) = \emptyset$
(5) $\mathsf{den}(Q)(g) = \mathsf{den}(p)(g)$, for all variable valuations $g \in G$, i.e., $\mathsf{den}(Q) = \mathsf{den}(p)$
(6) All bound variables occurring in Q, A, A_i are pairwise different. Thus, Q, A, A_i have no bound occurrences of q, p in them, i.e., Q, A, A_i have no recursion sub-terms that bind q, p (this condition can be skipped)
(7) $A_i\{ q :\equiv p \}$ is the *free replacement* of of q in A_i with p

The (chain) rule is:

$$C \equiv_{\mathsf{c}} [A_0 \text{ where } \{ q := Q, \ p := A, p_1 := A_1, \ \ldots, \ p_n := A_n \}] \tag{16a}$$
(chain)
$$\Rightarrow_{\mathsf{ch}} D \equiv_{\mathsf{c}} [A_0\{ q :\equiv p \} \text{ where } \{ p := A, p_1 := A_1\{ q :\equiv p \}, \\ \ldots, p_n := A_n\{ q :\equiv p \} \}] \tag{16b}$$

Often, we shall designate the extended sequences of reductions $A \Rightarrow_{\mathsf{ch}} B$ by $A \Rightarrow B$.

The chain rule introduced in [20] is a special case of the more general (chain) rule introduced here.

We designate the extended set of rules by $\mathsf{RedR_{ch}}$, by adding the rule (chain) to the set of the $\mathrm{L}^{\lambda}_{\mathrm{ar}}$-reduction rules (cong)–(λ), (γ^*), introduced in Sect. 3.

Definition 7 (Chain-Irreducible Terms). *We say that a term $A \in \mathsf{Terms}$ is* chain-irreducible *if and only if* (17) *holds:*

$$\textit{for all } B \in \mathsf{Terms}, \quad A \Rightarrow_{\mathsf{ch}} B \implies A \equiv_{\mathsf{c}} B \tag{17}$$

Note that a term A can be explicit, but reducible to a recursion term, as the examples in Sect. 4 demonstrate.

Chain Reduction Calculus The set of the reduction rules $\mathsf{RedR_{ch}}$ defines a *chain reduction* relation $\Rightarrow_{\mathsf{ch}}$ between terms, $A \Rightarrow_{\mathsf{ch}}{}^{*} B$, for all $A, B \in \mathsf{Terms}$. Usually, we write $A \Rightarrow_{\mathsf{ch}} B$, or simply $A \Rightarrow B$, instead of $A \Rightarrow_{\mathsf{ch}}{}^{*} B$.

Lemma 1. *Assume that, for appropriate types, $i \in \{0, 1, \ldots, n\}$ $(n \geq 0)$, the terms $Q, A, A_i \in \mathsf{Terms}$, and $q, p, p_i \in \mathsf{RecVars}$, satisfy the same conditions as in the* (chain) *rule above, and that, for given $y_j \in \mathsf{PureVars}$, for $j \in \{0, 1, \ldots, m\}$ $(m \geq 0)$, the term Q is of the form* (18)*:*

$$Q \equiv \lambda(\overrightarrow{y})(p(\overrightarrow{y})) \tag{18}$$

Then, the reduction (19a)–(19b) *holds:*

$$C \equiv_{\mathsf{c}} [A_0 \;\mathsf{where}\; \{\, q := \lambda(\overrightarrow{y})(p(\overrightarrow{y})),\; p := A, \overrightarrow{p} := \overrightarrow{A} \,\}] \tag{19a}$$

$$\Rightarrow_{\mathsf{ch}} D \equiv_{\mathsf{c}} [A_0\{\, q :\equiv p \,\} \;\mathsf{where}\; \{\, p := A, \overrightarrow{p} := \overrightarrow{A\{\, q :\equiv p \,\}} \,\}] \tag{19b}$$

Proof. The reduction (19a)–(19b) follows from the (chain) rule, for any term Q of the form (18), by (20):

$$\mathsf{den}(Q)(g) = \mathsf{den}\Big(\lambda(\overrightarrow{y})(p(\overrightarrow{y}))\Big)(g) = \mathsf{den}(p)(g), \text{for every } g \in G \tag{20}$$

The denotational equality (20) (and other similar and more complex recursion terms) is investigated in [16]. □ □

Lemma 2. *Assume that, for appropriate types, $i \in \{0, 1, \ldots, n\}$ $(n \geq 0)$, the terms $Q, A, A_i \in \mathsf{Terms}$, and $q, p, p_i \in \mathsf{RecVars}$, satisfy the conditions as in the* (chain) *rule* *above, and that, for given* $x, y_j \in \mathsf{PureVars}$, *for* $j \in \{0, 1, \dots, m\}$ $(m \geq 0)$, *the term* Q *is of the form* (21)*:*

$$Q \equiv \Big[\lambda(x)(x)\Big]\Big(\lambda(\overrightarrow{y})(p(\overrightarrow{y}))\Big) \tag{21}$$

Then, the reduction (22a)–(22b) *holds:*

$$C \equiv_{\mathsf{c}} [A_0 \text{ where } \{\, q := Q,\ p := A, \overrightarrow{p} := \overrightarrow{A}\,\}] \tag{22a}$$

$$\Rightarrow_{\mathsf{ch}} D \equiv_{\mathsf{c}} [A_0\{\, q :\equiv p\,\} \text{ where } \{\, p := A, \overrightarrow{p} := \overrightarrow{A\{\, q :\equiv p\,\}}\,\}] \tag{22b}$$

Proof. By Def. 3 of immediate terms and Def. 7 of chain irreducibility, it follows that the term Q in (21) is proper, explicit, irreducible. The reduction (22a)–(22b) follows from the (chain) rule. for any term Q of the form (21), by the denotational equality (23):

$$\mathsf{den}(Q) = \mathsf{den}\Big(\Big[\lambda(x)(x)\Big]\Big(\lambda(\overrightarrow{y})(p(\overrightarrow{y}))\Big)\Big) = \mathsf{den}(p) \tag{23}$$

□ □

Theorem 1 (Existence and Uniqueness of the Chain-Free Canonical Form). *For every* $A \in \mathsf{Terms}$, *there exist explicit, irreducible terms* $A_0, \dots, A_n$ $(n \geq 0)$, *such that the term* $\mathsf{cf}_{\mathsf{ch}}(A)$ *of the form* (24) *is chain irreducible:*

$$\mathsf{cf}_{\mathsf{ch}}(A) \equiv A_0 \text{ where } \{\, p_1 := A_1, \dots, p_n := A_n\,\} \tag{24}$$

and the following clauses hold:

(1) For every $\mathsf{c} \in \mathsf{Consts}$, c *occurs in* A *if and only if* c *occurs in* $\mathsf{cf}_{\mathsf{ch}}(A)$, *i.e.,* $\mathsf{Consts}(A) = \mathsf{Consts}(\mathsf{cf}_{\mathsf{ch}}(A))$

(2) $\mathsf{FreeVars}(A) = \mathsf{FreeVars}(\mathsf{cf}_{\mathsf{ch}}(A))$

(3) $A \Rightarrow_{\mathsf{ch}} \mathsf{cf}_{\mathsf{ch}}(A)$

(4) If A *is chain irreducible, then* $\mathsf{cf}_{\mathsf{ch}}(A) \equiv_{\mathsf{c}} A$

(5) (Uniqueness of the chain-free canonical forms up to congruence) If $A \Rightarrow_{\mathsf{ch}} B$, *then* $\mathsf{cf}_{\mathsf{ch}}(A) \equiv_{\mathsf{c}} \mathsf{cf}_{\mathsf{ch}}(B)$

(6) If $A \Rightarrow_{\mathsf{ch}} B$ *and* B *is chain irreducible, then* $B \equiv_{\mathsf{c}} \mathsf{cf}_{\mathsf{ch}}(A)$, *i.e.,* $\mathsf{cf}_{\mathsf{ch}}(A)$ *is a unique, up to congruence, irreducible term*

(7) If $A \Rightarrow B$ *(in the original reduction calculus of* $\mathrm{L}^{\lambda}_{\mathrm{ar}}$*), then* $\mathsf{cf}_{\mathsf{ch}}(A) \equiv_{\mathsf{c}} \mathsf{cf}_{\mathsf{ch}}(B)$

Proof. (1)–(3) is proved by structural induction on formation of terms and using the $\Rightarrow_{\mathsf{ch}}$-reduction rules; (4) by induction on the definition of the reduction relation; (5) follows from (3) and (4). The other statements can be proved similarly. □

Example 5 (A More Efficient Option for Rendering into Canonical Form). *A noun phrase, NP, e.g., "the cube", can be rendered directly into a term in a canonical form, from the canonical terms rendering the components of the NP.*

$$the \xrightarrow{\mathsf{render}} d \text{ where } \{ i := d(c),\, d := the \} \;: ((\tilde{\mathsf{e}} \to \tilde{\mathsf{t}}) \to \tilde{\mathsf{e}}) \tag{25a}$$

$$[the\ cube]_{NP} \xrightarrow{\mathsf{render}} T^0 \equiv i \text{ where } \{ i := d(c),\, d := the,\, c := cube \} \;: \tilde{\mathsf{e}} \tag{25b}$$

$$[is\ large]_{VP} \xrightarrow{\mathsf{render}} T_{isLarge} \equiv b \text{ where } \{ b := isLarge \} \;: (\tilde{\mathsf{e}} \to \tilde{\mathsf{t}}) \tag{25c}$$

The composition of the components can be rendered directly into the canonical form T^2, *as in* (26c), *from the canonical forms of the components in* (26a)–(26b), *by using* (chain) *rule and* $T_{isLarge}(T^0) \Rightarrow T^1 \Rightarrow_{\mathsf{ch}} T^2$*:*

$$\{ [The\ cube]_{NP}, [is\ large]_{VP} \}_S \xrightarrow{\mathsf{render}} T^2 \tag{26a}$$

$$\begin{aligned} T^1 \equiv b(e) \text{ where } \{ & e := i,\, i := d(c),\, d := the,\, c := cube, \\ & b := isLarge \} \end{aligned} \tag{26b}$$

$$\begin{aligned} T^1 \Rightarrow_{\mathsf{ch}} T^2 \equiv b(i) \text{ where } \{ & i := d(c),\, d := the,\, c := cube, \\ & b := isLarge \} \qquad \textit{by (chain) rule} \end{aligned} \tag{26c}$$

Note that the term T^1 is derived more efficiently in (26a)–(26b) than in (15a)–(15h), directly from the terms in canonical forms $T^0 \equiv \mathsf{cf}(T^0)$ and $T_{isLarge} \equiv \mathsf{cf}(T_{isLarge})$. Then, (26c) removes the chain assignments $e := i$, $i := d(c)$, by the (chain) rule.

Efficiency by Underspecification Furthermore, the underspecified term (27) represents a wide class of predications $b(i)$ of properties $\mathsf{den}(b)$, which are designated by the unspecified, free memory variable $b \in \mathsf{RecV}_{(\tilde{\mathsf{e}} \to \tilde{\mathsf{t}})}$, to individual entities $\mathsf{den}(i)$ designated by the definite descriptions $d(c) : \tilde{\mathsf{e}}$. Each $i := d(c)$ has the unspecified property $c \in \mathsf{RecV}_{(\tilde{\mathsf{e}} \to \tilde{\mathsf{t}})}$ that can be instantiated by a specific property, e.g., $c := cube$, as in (26a)–(26c). Specified, chain-free, canonical forms are obtained by specification assignments to the memory locations b and c, e.g., as for T^2 from (27):

$$T \equiv b(i) \text{ where } \{ i := d(c),\, d := the \} \tag{27}$$

Theorem 2 (Effective Reduction Calculi). *For every term* $A \in \mathsf{Terms}$, *its chain-free canonical form* $\mathsf{cf}_{\mathsf{ch}}(A)$ *is effectively computed, by the chain reduction calculus, i.e.:* $A \Rightarrow \mathsf{cf}_{\mathsf{ch}}(A)$.

The proof is by structural induction on $A \in \mathsf{Terms}$ using Theorem 1. The full, detailed proof is long and outside of the topic of this paper.

6 Conclusion and Future Work

The formal languages of acyclic recursion $\mathrm{L}^{\lambda}_{\mathrm{ar}}$ and of full recursion L^{λ}_{r} provide a type-theoretic formalization of the mathematical notion of algorithm. The canonical form $\mathsf{cf}(A)$ of each proper term A determines the algorithm for computing the value $\mathsf{den}(A)$ of the recursive function denoted by A, which is encoded by the respective one-argument function, via currying technique. That is, the algorithmic semantics of a proper (i.e., non-immediate) term A, is determined by its canonical form $\mathsf{cf}(A)$.

In this paper, I have extended the reduction calculus of $\mathrm{L}^{\lambda}_{\mathrm{ar}}$ introduced by Moschovakis [25], by adding the (chain) rule. The goal is to reduce the complexity of reductions in $\mathrm{L}^{\lambda}_{\mathrm{ar}}$. The extended (chain) rule eliminates assignments of the form $q := Q, p := A$, for $Q, A \in \mathsf{Terms}$, Q explicit, irreducible, and memory locations $p, q \in \mathsf{RecVars}$, such that $\mathsf{den}(Q) = \mathsf{den}(p)$. The assignment $q := Q$ in such a pair correspond to computations of the value $\mathsf{den}(Q)$, which is identical to that in the memory location p. In cases when the term A is proper, i.e., non-immediate, the value saved in the memory slot p, i.e., $\mathsf{den}(p)$, is computed via the assignment $p := A$.

For broader and more useful applications of TTAR, $\mathrm{L}^{\lambda}_{\mathrm{ar}}$, we have included also the (γ^*) rule, introduced in [12], to reduce unnecessary, vacuous λ-abstractions.

The special cases of chain assignments, were introduced in [20], for $q := \lambda(\overrightarrow{y})(p(\overrightarrow{y}))$, p A, i.e., a "copy" of the value of A is fetched from p and saved in q. By the introduced rule (chain), $q := \lambda(\overrightarrow{y})(p(\overrightarrow{y}))$ is removed and all the occurrences of q are replaced with p. Thus, the chain free canonical forms of terms determine more efficient algorithms for computing their denotations.

Chain assignments can arise during syntax-semantics interface in compositional syntax-semantics rules, which put together syntactic components in larger parses, by linking the semantic rendering of the smaller parse components, saved in memory locations, see e.g. [10]. The algorithmic semantics of *proper* (i.e., non-immediate) terms A, is determined by their canonical forms $\mathsf{cf}(A)$:

$$\mathsf{alg}(A) = \mathsf{alg}(\mathsf{cf}(A)) \text{ computes } \mathsf{den}(A) = \mathsf{den}(\mathsf{cf}(A)) \tag{28}$$

$$\underbrace{\textsf{Syntax of } \mathrm{L}^{\lambda}_{\mathrm{ar}}\ (\mathrm{L}^{\lambda}_{r}) \Longrightarrow \textsf{Algorithms: } \mathsf{alg}(A) = \mathsf{alg}(\mathsf{cf}(A)) \Longrightarrow \textsf{Denotations } \mathsf{den}(A)}_{\textsf{Algorithmic Syntax-Semantics Interface of } \mathrm{L}^{\lambda}_{\mathrm{ar}}(\mathrm{L}^{\lambda}_{r})}$$

Note that syntax-semantic compositions of the smaller components into larger ones, by linking them, can introduce chain assignments, including by the (ap) rule, e.g., when he head part of B is immediate. The introduced reduction rule (chain) removes them and results in more efficient algorithms. This is demonstrated by Examples 3, 4, 5. The type theory L^{λ}_{ar} provides rich varieties of terms available for compositions depending on the choices of computational grammars of formal and natural languages, including large-scale grammars, see [8, 10]. The introduced type theory of algorithms for chain-free calculations is new. Extended studies of more properties and applications are forthcoming.

References

[1] Copestake, A., Flickinger, D., Pollard, C., Sag, I.: Minimal recursion semantics: an introduction. Research on Language and Computation **3**, 281–332 (2005), https://doi.org/10.1007/s11168-006-6327-9

[2] Gallin, D.: Intensional and Higher-Order Modal Logic: With Applications to Montague Semantics. North-Holland Publishing Company, Amsterdam and Oxford, and American Elsevier Publishing Company (1975), https://doi.org/10.2307/2271880

[3] Loukanova, R.: Russellian and Strawsonian Definite Descriptions in Situation Semantics. In: Gelbukh, A. (ed.) Computational Linguistics and Intelligent Text Processing. CICLing 2001, Lecture Notes in Computer Science, vol. 2004, pp. 69–79. Springer, Berlin, Heidelberg (2001), https://doi.org/10.1007/3-540-44686-9_6

[4] Loukanova, R.: Quantification and Intensionality in Situation Semantics. In: Gelbukh, A. (ed.) Computational Linguistics and Intelligent Text Processing. CICLing 2002, Lecture Notes in Computer Science, vol. 2276, pp. 32–45. Springer, Berlin, Heidelberg (2002), https://doi.org/10.1007/3-540-45715-1_3

[5] Loukanova, R.: Situation Theory, Situated Information, and Situated Agents. In: Nguyen, N.T., Kowalczyk, R., Fred, A., Joaquim, F. (eds.) Transactions on Computational Collective Intelligence XVII, Lecture Notes in Computer Science, vol. 8790, pp. 145–170. Springer, Berlin, Heidelberg (2014), https://doi.org/10.1007/978-3-662-44994-3_8

[6] Loukanova, R.: Acyclic Recursion with Polymorphic Types and Underspecification. In: van den Herik, J., Filipe, J. (eds.) Proceedings of the 8th International Conference on Agents and Artificial Intelligence. vol. 2, pp. 392–399. SciTePress — Science and Technology Publications, Lda. (2016). https://doi.org/10.5220/0005749003920399

[7] Loukanova, R.: Relationships between Specified and Underspecified Quantification by the Theory of Acyclic Recursion. ADCAIJ: Advances in Distributed Computing and Artificial Intelligence Journal **5**(4), 19–42 (2016), https://doi.org/10.14201/ADCAIJ2016541942

[8] Loukanova, R.: An Approach to Functional Formal Models of Constraint-Based Lexicalized Grammar (CBLG). Fundamenta Informaticae **152**(4), 341–372 (2017). https://doi.org/10.3233/FI-2017-1524

[9] Loukanova, R.: Typed Theory of Situated Information and its Application to Syntax-Semantics of Human Language. In: Christiansen, H., Jiménez-López, M.D., Loukanova,

R., Moss, L.S. (eds.) Partiality and Underspecification in Information, Languages, and Knowledge, pp. 151–188. Cambridge Scholars Publishing (2017)

[10] Loukanova, R.: Computational Syntax-Semantics Interface with Type-Theory of Acyclic Recursion for Underspecified Semantics. In: Osswald, R., Retoré, C., Sutton, P. (eds.) IWCS 2019 Workshop on Computing Semantics with Types, Frames and Related Structures. Proceedings of the Workshop. pp. 37–48. The Association for Computational Linguistics (ACL), Gothenburg, Sweden (Jun 2019), https://www.aclweb.org/anthology/W19-1005

[11] Loukanova, R.: Gamma-Reduction in Type Theory of Acyclic Recursion. Fundamenta Informaticae **170**(4), 367–411 (2019), https://doi.org/10.3233/FI-2019-1867

[12] Loukanova, R.: Gamma-Star Canonical Forms in the Type-Theory of Acyclic Algorithms. In: van den Herik, J., Rocha, A.P. (eds.) Agents and Artificial Intelligence. ICAART 2018, Lecture Notes in Computer Science, book series LNAI, vol. 11352, pp. 383–407. Springer International Publishing, Cham (2019), https://doi.org/10.1007/978-3-030-05453-3_18

[13] Loukanova, R.: Type-Theory of Acyclic Algorithms for Models of Consecutive Binding of Functional Neuro-Receptors. In: Grabowski, A., Loukanova, R., Schwarzweller, C. (eds.) AI Aspects in Reasoning, Languages, and Computation, vol. 889, pp. 1–48. Springer International Publishing, Cham (2020), https://doi.org/10.1007/978-3-030-41425-2_1

[14] Loukanova, R.: Currying Order and Restricted Algorithmic Beta-Conversion in Type Theory of Acyclic Recursion. In: Materna, P., Jespersen, B. (eds.) Logically Speaking. A Festschrift for Marie Duží. Book Tribute, vol. 49, p. 285–310. College Publications (12 2022). https://doi.org/10.13140/RG.2.2.34553.75365

[15] Loukanova, R.: Algorithmic Dependent-Type Theory of Situated Information and Context Assessments. In: Omatu, S., Mehmood, R., Sitek, P., Cicerone, S., Rodríguez, S. (eds.) Distributed Computing and Artificial Intelligence, 19th International Conference. vol. 583, pp. 31–41. Springer International Publishing, Cham (2023), https://doi.org/10.1007/978-3-031-20859-1_4

[16] Loukanova, R.: Eta-Reduction in Type-Theory of Acyclic Recursion. ADCAIJ: Advances in Distributed Computing and Artificial Intelligence Journal **12**(1), 1–22, e29199 (2023), https://doi.org/10.14201/adcaij.29199

[17] Loukanova, R.: Logic Operators and Quantifiers in Type-Theory of Algorithms. In: Bekki, D., Mineshima, K., McCready, E. (eds.) Logic and Engineering of Natural Language Semantics. LENLS 2022, Lecture Notes in Computer Science (LNCS), vol. 14213, pp. 173–198. Springer Nature Switzerland, Cham (2023), https://doi.org/10.1007/978-3-031-43977-3_11

[18] Loukanova, R.: Restricted Computations and Parameters in Type-Theory of Acyclic Recursion. ADCAIJ: Advances in Distributed Computing and Artificial Intelligence Journal **12**(1), 1–40 (2023), https://doi.org/10.14201/adcaij.29081

[19] Loukanova, R.: Semantics of Propositional Attitudes in Type-Theory of Algorithms. In: Bekki, D., Mineshima, K., McCready, E. (eds.) Logic and Engineering of Natural

Language Semantics. LENLS 2023, Lecture Notes in Computer Science (LNCS), vol. 14569, pp. 260–284. Springer Nature Switzerland AG, Cham (2024), https://doi.org/10.1007/978-3-031-60878-0_15

[20] Loukanova, R.: Type-theory of algorithms with chain-free memory. In: Chinthaginjala, R., Sitek, P., Min-Allah, N., Matsui, K., Ossowski, S., Rodríguez, S. (eds.) Distributed Computing and Artificial Intelligence, 21st International Conference (DCAI 2024). Lecture Notes in Networks and Systems, vol. 1259, pp. 65–76. Springer Nature Switzerland, Springer, Cham (2025), https://doi.org/10.1007/978-3-031-82073-1_7

[21] Montague, R.: The Proper Treatment of Quantification in Ordinary English. In: Hintikka, J., Moravcsik, J., Suppes, P. (eds.) Approaches to Natural Language, vol. 49, pp. 221–242. Synthese Library. Springer, Dordrecht (1973), https://doi.org/10.1007/978-94-010-2506-5_10

[22] Moschovakis, Y.N.: The formal language of recursion. Journal of Symbolic Logic **54**(4), 1216–1252 (1989), https://doi.org/10.1017/S0022481200041086

[23] Moschovakis, Y.N.: Sense and denotation as algorithm and value. In: Oikkonen, J., Väänänen, J. (eds.) Logic Colloquium '90: ASL Summer Meeting in Helsinki, Lecture Notes in Logic, vol. Volume 2, pp. 210–249. Springer-Verlag, Berlin (1993), https://projecteuclid.org/euclid.lnl/1235423715

[24] Moschovakis, Y.N.: The logic of functional recursion. In: Dalla Chiara, M.L., Doets, K., Mundici, D., van Benthem, J. (eds.) Logic and Scientific Methods, vol. 259, pp. 179–207. Springer, Dordrecht (1997), https://doi.org/10.1007/978-94-017-0487-8_10

[25] Moschovakis, Y.N.: A Logical Calculus of Meaning and Synonymy. Linguistics and Philosophy **29**(1), 27–89 (2006), https://doi.org/10.1007/s10988-005-6920-7

[26] Müller, S., Abeillé, A., Borsley, R.D., Koenig, J.P. (eds.): Head-Driven Phrase Structure Grammar. No. 9 in Empirically Oriented Theoretical Morphology and Syntax, Language Science Press, Berlin (2024). https://doi.org/10.5281/zenodo.13637708

[27] Plotkin, G.D.: LCF considered as a programming language. Theoretical Computer Science **5**(3), 223–255 (1977), https://doi.org/10.1016/0304-3975(77)90044-5

[28] Sag, I.A., Wasow, T., Bender, E.M.: Syntactic Theory: A Formal Introduction. CSLI Publications, Stanford, California (2003)

[29] Scott, D.S.: A type-theoretical alternative to ISWIM, CUCH, OWHY. Theoretical Computer Science **121**(1), 411–440 (1993), https://doi.org/10.1016/0304-3975(93)90095-B

Received 17 March 2025

Generative Artificial Intelligence for Controlled Content Generation

J. M. Torres, C. A. Martin, M. A. Bacallado, R. Aguilar, S. Alayón
Department of Computer Engineering and Systems,
University of La Laguna, Spain
{jmtorres,camartin,mbacalll,raguilar,salayon}@ull.edu.es

Abstract

The main objective of this article is to provide an overview of how Generative Artificial Intelligence techniques can be used to create content in a controlled manner for applications that require high reliability and accuracy. Generative Artificial Intelligence, capable of generating various types of content, such as text, images, video, and audio, has revolutionised many areas of knowledge, but it must be considered that these systems can suffer from 'hallucinations' and produce false or inconsistent information. For this reason, having a system that allows their controlled use in sensitive contexts is of vital importance. This paper presents a methodology that allows the user to use a Generative Artificial Intelligence system to create new content according to the user's needs based on controlled and user-specified sources. The operation of the proposed system is illustrated by means of a concrete example of application in education, the automatic generation of test-type exams.

Keywords: Generative Artificial Intelligence, Retrieval Augmented Generation, User-centred automation, Education, Exam generation

1 Introduction

Artificial Intelligence (AI) is one of today's most disruptive and transformative technologies. The great potential of AI programs to learn to perform tasks like people in multiple contexts has triggered its application in all fields of knowledge.

The goal of designing machines with intelligence and autonomy of their own emerged in the 1950s [1], with the appearance of the first neural network, the perceptron [2]. The neural network theory developed over the next 20 years, giving rise

This work was funded by CajaCanarias Foundation and La Caixa Foundation (grant number 2023DIG11).

to multilayer networks trained with backpropagation algorithms, which were the basis of the first intelligent systems [3]. At the beginning of this century, important advances were made in speech recognition, image processing, autonomous driving, and game development. The first Natural Language Processing (NLP) systems emerged, with the aim of enabling computers to understand human language [4]. The 2010s are marked by the development of artificial intelligence systems based on Deep Learning, a machine learning technique that uses complex artificial neural networks to mimic the way the human brain processes information. These networks can analyse large amounts of data to detect patterns and make predictions, which has led to major advances in areas such as voice and image recognition, machine translation, natural language processing, among others [5], until becoming a Generative AI (GAI), i.e. an AI capable of generating new content in the form of text, audio, images, or video [6]. It is precisely the use of this last evolution of AI that is studied in this article.

GAI is currently revolutionising the way we work and communicate; the widespread use of GPT-4 [7], Copilot [8] or Gemini [9] are clear examples of this. GAI systems are based on the use of Large Language Models (LLM). LLMs are composed of transformers, which are models that incorporate *attention mechanisms* to decide which words are the most relevant in a text to analyse context and meaning in a way similar to that of humans [10]. These models have a very high internal complexity, with many layers, and hundreds of billions of internal parameters or more, which must be tuned with a huge amount of text data [11]. LLMs have a deep understanding of language, human-like text generation capabilities, contextual awareness, and strong problem-solving skills, making them invaluable in a variety of domains: search engines, translation, content and product generation, chatbots, etc. [12].

However, these models also have important limitations. They sometimes provide incorrect answers and/or generate incorrect information (*hallucinations*), may be overly sensitive to arbitrary differences in question wording (*prompt*), and may not be able to systematically clarify ambiguous questions, as the ChatGPT authors themselves point out in [7]. Despite their advanced technology, these models sometimes fail to understand the context and generate text with falsehoods that are difficult to detect, since their answer seems convincing and consistent with what a human could answer.

To solve this problem, Retrieval Augmented Generation (RAG) can be applied [13]. This technique redirects the LLM to retrieve relevant information from predetermined knowledge sources. In this way, the accuracy, relevance, and timeliness of the answers generated by the model are improved [14]. Figure 1 shows schematically how queries are performed in this type of system. The left-hand side

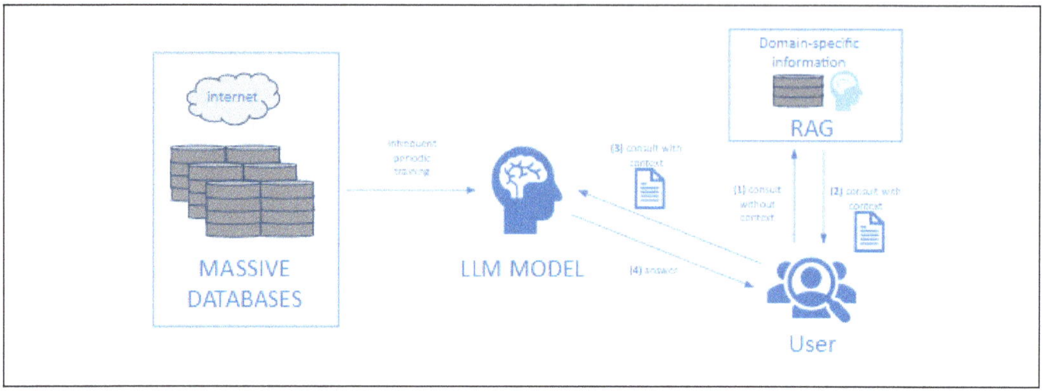

Figure 1: Schematic diagram of how the interaction between RAG and LLM can improve generative results.

highlights the fact that the LLM model has been trained on massive datasets obtained from large information systems or the Internet. The process starts when the user wants to perform an initial query. However, the LLM model needs an appropriate context to construct an accurate answer. For this reason, the user sends the query to the RAG system to retrieve domain-specific information from external data sources. The RAG system returns the query to the user enriched with specific information, providing a concrete and up-to-date context. This enriched query is transferred to the Language Model (LLM). The coherent and generated answer with relevant information is sent to the user, reducing the risk that it contains outdated, inaccurate, or in some cases incorrect information due to hallucination.

The aim of this paper is to explain how a user can easily implement a RAG system for personal applications. In some applications, ensuring the reliability of the responses of the generative AI model is very important due to the implications of handling incorrect or biased information in these contexts. For this reason, the performance of the proposed system will be tested with a case of use in the highly sensitive field of education. In this case of use, the user is a teacher and will be shown how to use the proposed RAG system to design an exam generation tool. This tool will operate as a "virtual teacher assistant" for the automatic generation of assessment tests based on specific material. The teacher will be able to propose to the assistant a topic or aspect of the material provided on which he/she wants to ask questions, as well as the difficulty of the questions to be generated. With this information, the assistant will provide multiple-choice questions with possible alternative answers, indicating which of them is the correct one. The architecture of the proposed system will be explained in detail in this example.

In Spain, in the university environment, teaching staff must primarily perform teaching and research tasks and sometimes also management tasks [15]. One of the teaching tasks teachers must perform is the assessment of their students, and the most widely used method of assessment are exams [16, 17, 18]. The process of properly designing these tests is very costly in terms of the time invested by the teacher [19]. For this reason, the application of the proposed system to this problem seems useful: an automated method of generating exams can allow the teacher more time for other tasks.

As manual creation of questions and manual evaluation of answers is labour intensive and time consuming, many teachers are interested in developing tools that automate this process [20]. In fact, in [21], it is highlighted that, for the correct integration of GAI in the teaching activity, it would be important 1) to perform assessment tasks through GAI to check to what extent the tool can perform the task and 2) to make students use GAI as part of the assessment process. However, there is little in the literature that addresses this issue from a practical point of view. One of them is presented in [22], where the authors share a web application that allows teachers to use an interface that combines Large Language Models (LLM), such as ChatGPT, with options to specify questions of interest, sample solutions, and assessment instructions for automated scoring of answers. This tool, tested with undergraduate and postgraduate natural science students, is focused on automating the scoring of extended text-based responses according to the sample solutions and scoring instructions provided.

In the following sections, the full functionality of the proposed RAG system, which acts as a virtual assistant for the teacher, will be explained. The operation of each of its parts and the potential advantages of its use will be shown. Finally, the results and conclusions obtained will be discussed, leaving open the possibility of adaptations to other domains and new developments.

2 Proposed Method

2.1 Overview of the Proposed RAG System

The designed system is composed of two subsystems: an *information extraction module*, and a *generator module*. Figure 2 presents a conceptual diagram of both subsystems and how they interact with each other and with the teacher (user of the tool).

The way the system works is as follows: the topic on which questions are to be generated is indicated in the Exam Generation Module (step 1 - *topic*), the teacher must also indicate to the Information Extraction Module which are the appropriate resource sources to extract the documents containing the information on that topic (step 2 - *select source*). The job of the information extraction module will be to extract the information from the indicated sources and supply the exam generator module with the documents (step 3 - *documents*), so that using AI it can generate the questions and alternative answers for the teacher (step 4 - *question, alternative answers, correct answer*). In the following subsections, each of the modules that make up the system will be developed in more detail.

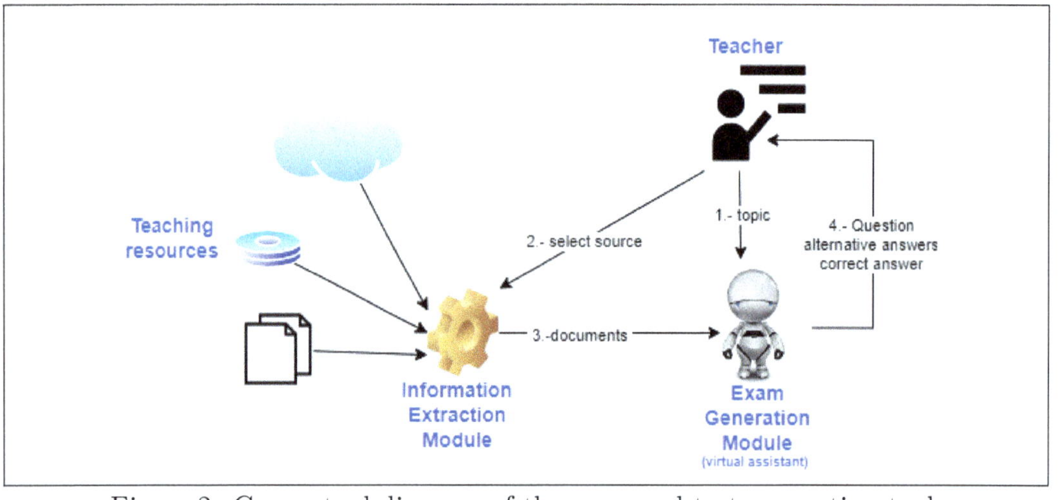

Figure 2: Conceptual diagram of the proposed test generation tool.

2.2 Information Extraction Module

The information extraction module is responsible for analysing the available teaching sources and extracting documents related to a specific topic. The user of the system must provide the teaching sources he/she considers appropriate for the tackled topic. The teaching material must be in a digital format. Today's teachers usually have different sources of materials that they use as notes or reference documentation to teach their students. Typically, these are hosted on websites, virtual teaching platforms, or file repositories accessible by students. Any of these resources would be valid for the information extractor module.

Separating the information extraction functionality from the exam generation in a separate module has a great advantage: it allows the tool to be adapted to other environments with learning materials stored in other digital formats. It is important to mention that the complexity of the information extraction module can grow significantly if the learning materials are poorly formatted. For example, paragraph-separated text documents and structured documents based on markup tags (*html* or *markdown*) are materials that do not require complex extraction modules. On the other hand, teaching materials based on scanned images, audio files or video files, or documents without hierarchical content structure are the types of material that require the use of complex extractors.

In this work, we have used as a source of teaching material a website where the notes of a subject of Operating Systems corresponding to the second year of the university degree in Computer Engineering at the University of La Laguna are hosted in HTML format. This material can be consulted at [23].

The information extraction module is based on a Python development that makes use of functions from the following external libraries: Beautiful Soup 4 [24], for extracting data from HTML files, Haystack [25] for building LLM applications, and Markdownify [26] for converting text to Markdown format.

The functions implemented in the extraction module developed in this work are shown schematically in Figure 3. As can be seen, there is an input function called *extract_website_documents* that receives a list of web addresses (URLs) as a source of information (teaching material on the subject) and returns a list of documents, in which each document is a different section of the supplied addresses converted to plain text. For this purpose, the function *download_website_sections* is used, which returns sections in markdown format. These sections are then converted to plain text using the *MarkdownConverter* object from the Haystack library.

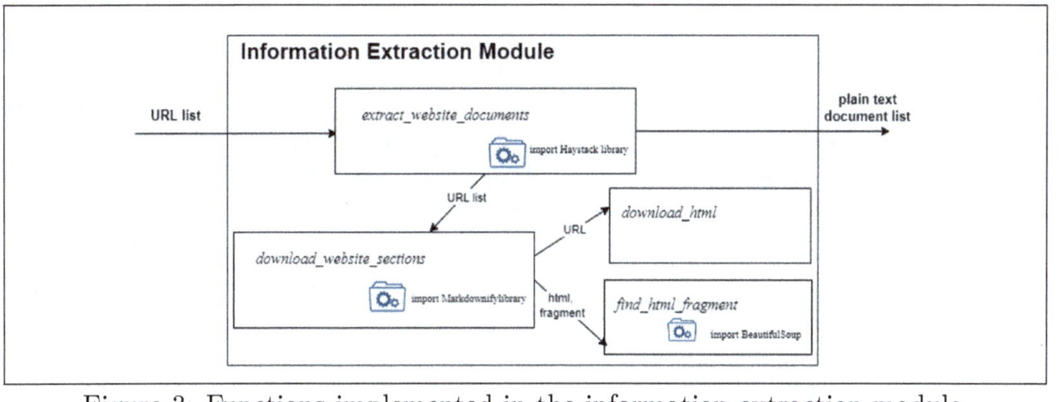

Figure 3: Functions implemented in the information extraction module.

2.3 Generator Module

The generator module is in charge of generating multiple-choice questions on a topic proposed by the system user. This module is based on the use of a RAG system capable of generating new content from the knowledge provided by the information extraction module, specifically an LLM. To implement the exam generator module proposed in this article, it is sufficient to select one of the LLMs already available on the market. The only requirement is to check that the chosen version is able to generate questions (and possible answers to these questions) with an acceptable prompt length, that they are logical and well written according to the criteria of the teacher user of the tool.

The *prompt* refers to the instruction provided to an AI system to perform a specific task or respond to a query. The quality of the prompt is essential, as it determines the accuracy and relevance of the responses generated by the AI [27]. An effective prompt should include a clear explanation of the task you want the LLM to do and additional information to place the task in a specific framework. In this paper, the GPT4 model has been chosen as the RAG system for the test generator module.

The exam generator module uses the list of plain text documents received from the information extraction module. User-configurable parameters can be used to define features to adjust the model result: the name of the model (which in our case is *GPT-4*), the difficulty of the questions to be generated (*difficulty* parameter), the cognitive level (*cognitive_level* parameter), and the topic on which the question and possible answers will be generated (*query* parameter).

The module, using an object of the *PromptNode* class of the Haystack library, creates an instance that generates a specific prompt to ask the model a question about a specific topic with several alternative answers. To do this, it is necessary to use a template containing the following text:

"Given the context, please generate a multiple-choice question in Spanish on the indicated topic, with 5 possible answers, only one correct and tell which is the correct one. The question must be precise and appropriate for students of the subject operating systems in the second year of a computer engineering degree. It must be a question of the indicated cognitive level and difficulty. Regarding style, I want you to be formal and do not use the first or second person. Please, generate the question in GIFT format, to easily import it in Moodle .

Context: {join(documents)}

Topic: {query}

Difficulty: {difficulty}

CognitiveLevel: {cognitive_level}

Answer:"

The diagram in Figure 4 shows schematically the composition of the exam generator module and the use of the model. As can be seen in Figure 4, the input to the module is the list of plain text documents from the information extraction module, and the input parameters supplied by the user to generate the prompt (*topic, difficulty, cognitive level*). The output of the module will be the exam question (based on the topic entered as a parameter, using the supplied list of documents as materials, and with the indicated difficulty and cognitive level) together with the alternative answers. All this in GIFT (General Import Format Technology) format as indicated in the template.

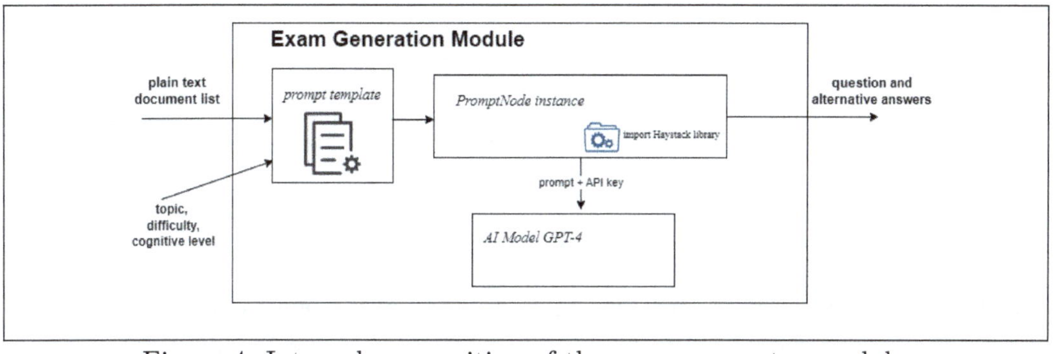

Figure 4: Internal composition of the exam generator module.

2.4 Full Operation of the System

The previous sections have shown the composition and operation of the information extraction and generation modules. The main program must be the part that combines the execution of both modules. In our work, it has been implemented in Python.

Once the plain text documents have been obtained from the information extractor, the main program must generate the prompt with the parameters collected from the user and execute the prompt method of the *PromptNode* object instance of the exam generation module. The result of the execution may be displayed on the console or saved on the drive for later import into a learning tool. The Google Collab cloud service has been used for the development and execution of the main program.

Figure 5 shows how the parameters can be supplied to the main programme that links the two modules and runs the script on the cloud service. This is the interface through which the user accesses the system. The code of the proposed system applied to the automatic generation of multiple-choice tests can be freely downloaded from [28].

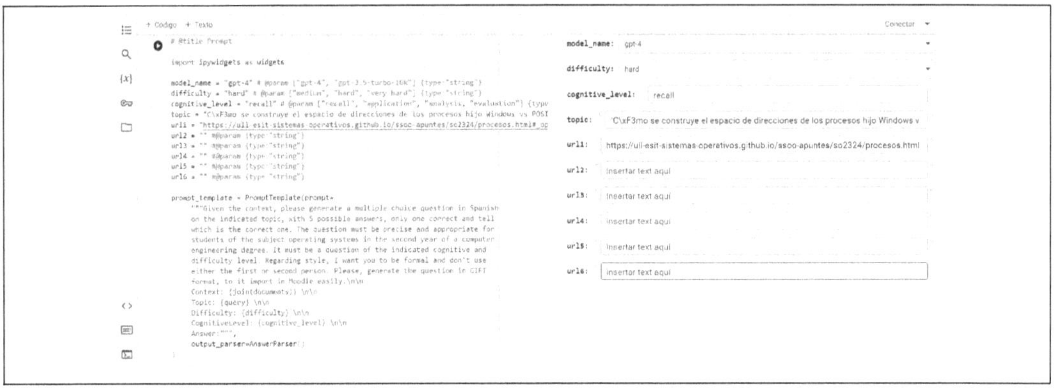

Figure 5: User interface of the exam generator system.

3 Results and Discussion

We have used the system described in section 2 on 50 topics extracted from the headings of the teaching materials specified for the subject of Operating Systems of the 2nd year of the Degree in Computer Engineering at the University of La Laguna. 245 possible questions were generated on the requested topics, with their 5 possible alternative answers in which only one was correct. From the total of the above questions (with their alternative answers), the subject teachers selected 40 questions, and for each of them 4 alternative answers (eliminating from the set of 5 one of the incorrect ones that was less clear or too obvious). The parameters *difficulty* and *cognitive level* were common to all questions, their values being *hard* and *recall.* Table 1 shows some of the questions generated with their alternative answers. All generated material was requested in GIFT format to facilitate its import into e-learning platforms, e.g., Moodle [29]. More examples of generated exams can be found at [28].

The proposed exam generator system has been used during the 2023-2024 and 2024-2025 school years in the Operating Systems subject with satisfactory results, so its use is planned for the following courses. The teachers involved in this subject have observed that AI-generated examinations are significantly more varied than manually generated ones, less ambiguous (the questions are clearer and more objective), and more complete (in the sense that they cover all the content of the subject, leaving no gaps). The number of students who have passed the subject in these courses, taking the exams generated by the AI system, has been similar to the number of passes in previous courses, where the exams were generated manually by the faculty. This shows that artificially generated exams are of similar difficulty.

The time savings for teachers are very significant. The latter can be estimated

Topic: "Planning algorithms"	
Question:	Which of the following statements about multilevel feed-forward queue scheduling algorithms is most accurate?
Answer A	All processes are assigned to a single queue but with different priority levels.
Answer B	The assignment of a process to a queue is done taking into account its estimated CPU burst time, to reduce waiting times.
Answer C	Multiple queues are used, and each queue must have a different scheduling algorithm to take advantage of its benefits.
Answer D	**Using a priority scheduler to choose the appropriate queue is common in modern operating systems.**
Topic: "System library"	
Question:	What is the main function of the system library?
Answer A	It provides assembly language instructions for system calls.
Answer B	It allows the standard library to directly access CPU instructions.
Answer C	**It is responsible for making the necessary system calls when programs invoke system library functions.**
Answer D	It controls access to operating system services, preventing programs from making system calls directly.

Table 1: Examples of questions and answers generated with different topics. The correct answer is marked in bold.

quantitatively. The volume of the teaching material used in this test is 113746 words, which is equivalent to about 228 pages (considering about 500 words per page). An adult is estimated to read about 200 words per minute, which means that a teacher would take approximately 9.5 hours to read the material. However, preparing an exam with various questions on the different topics of a teaching material is a job that requires much more effort than simply reading it. In addition to reading, there is the work of proposing alternative answers to different aspects of the material. Assuming a factor of 2.5 for the effort to prepare the questions with respect to reading the material, we could consider that preparing the tests for a subject would take the teacher approximately 24 hours (9.5 hours $\times$ 2.5 = 23.75 hours), which is equivalent to 3 full working days. These calculations more than justify the effort to automate this task and to systematise the methodology proposed in this article for as many subjects as possible.

4 Conclusions

This paper has shown how a RAG system can be implemented in a simple way and adapted to different contexts. The system is composed of two modules, one for information extraction and the other for content generation, to increase its flexibility.

The proposed system, in the application designed for education, has been converted into an automatic system for the generation of test-type questions. The system can be applied in the elaboration of exams in any area of knowledge or subject, if the teaching material is in digital format.

This tool has been tested in the evaluation of the subject Operating Systems of the Computer Engineering Degree of the University of La Laguna, proving to be useful and efficient. The automatic question generation system proposed in this work has proven to be an effective tool to significantly reduce teaching workload. The use of automation techniques and RAG has not only contributed to this reduction of effort but has also provided diversity of questions and adequacy to the specified level, improving the quality of the results, and allowing teachers to lighten more routine tasks and focus on higher quality teaching. It would be interesting to conduct a comparative study between the questions generated by the proposed methodology and other known content generators, such as Copilot. This will be investigated in future work.

An important aspect should be considered when transferring the results of this work to other use cases. In certain areas of knowledge, the materials are in formats that are not very suitable for the automatic generation of documents understandable by a generative AI. The digitisation and structured composition of these materials is a task prior to the use of a tool such as that presented. There are several techniques that can be used individually or in combination to solve this problem: Optical Character Recognition (OCR), Natural Language Processing (NLP), Machine Learning (ML) and Deep Learning (DL), Document Layout Analysis (DLA), and Robotic Process Automation (RPA).

The authors of this article are clear on the continuation of this work. The next step will be to study the applicability of the RAG system in another very sensitive environment, that of Justice. The choice of this context for the application is not accidental. The authors are currently immersed in a research project whose ultimate goal is the automation of routine tasks and the reduction of response times of the Commercial Courts, in order to boost the efficiency and quality of their services. Some of the tasks to be automated are routine and involve the search and draughting of documentation, which, with the help of an GAI system, could be quickly solved. Incorporating a RAG system such as the one described in this paper is essential to ensure the reliability of the responses of the generative system.

References

[1] J. Moor. The Dartmouth College Artificial Intelligence Conference: The Next Fifty Years. AI Magazine, 27:87-91, 2006.

[2] F. Rosenblatt. The perceptron: A probabilistic model for information storage and organization in the brain. Psychological Review, 65(6):386-408, 1958.

[3] H.M. Collins. Expert systems and the science of knowledge. In: The social construction of technological systems: New directions in the sociology and history of technology: Anniversary Edition, pages 321-320. MIT Press, Boston, 2012.

[4] E. Cambria and B. White. Jumping NLP Curves: A Review of Natural Language Processing Research. IEEE Computational Intelligence Magazine, 9(2):48-57, 2014.

[5] R. Wason. Deep learning: Evolution and expansion. Cognitive Systems Research, 52:701-708, 2018.

[6] S. Feuerriegel, J. Hartmann, C. Janiesch, and P. Zschech. Generative AI. Business & Information Systems Engineering, 66(1):111-126, 2024.

[7] OpenAI. GPT-4 Technical Report. Available at: https://openai.com/gpt-4, 2024.

[8] Microsoft Corporation. Microsoft Copilot. Available at: https://www.microsoft.com/es-es/microsoft-copilot, 2024.

[9] Google. Gemini AI. Available at: https://gemini.google.com/?hl=es-ES, 2024.

[10] A. Vaswani, N. Shazeer, N. Parmar, J. Uszkoreit, L. Jones, A.N. Gomez, L. Kaiser, and I. Polosukhin. Attention is all you need. ArXiv preprint 1706.03762, 2017.

[11] W.X. Zhao, K. Zhou, J. Li, T. Tang, X. Wang, Y. Hou, Y. Min, B. Zhang, J. Zhang, Z. Dong, Y. Du, C. Yang, Y. Chen, Z. Chen, J. Jiang, R. Ren, Y. Li, X. Tang, Z. Liu, P. Liu, J.-Y. Nie, and J.-R. Wen. A survey of large language models. ArXiv preprint 2023.18223, 2023.

[12] B. Almarie, P.E.P. Teixeira, K. Pacheco-Barrios, C.A. Rossetti, and F. Fregni. Editorial - The Use of Large Language Models in Science: Opportunities and Challenges. Principles and Practice of Clinical Research, 9(1):1-4, 2023.

[13] P. Lewis, E. Perez, A. Piktus, F. Petroni, V. Karpukhin, N. Goyal, H. Kuttler, M. Lewis, W.T. Yih, T. Rocktaschel, S. Riedel, and D. Kiela. Retrieval-Augmented Generation for Knowledge-Intensive NLP Tasks. In: H. Larochelle, M. Ranzato, R. Hadsell, M.F. Balcan, and H. Lin, editors, Advances in Neural Information Processing Systems, volume 33, pages 9459-9474. Curran Associates, Inc., 2020.

[14] Y. Gao, Y. Xiong, X. Gao, K. Jia, J. Pan, Y. Bi, Y. Dai, J. Sun, Q. Guo, M. Wang, and H. Wang. Retrieval-Augmented Generation for Large Language Models: A Survey. ArXiv preprint 2312.10997, 2023.

[15] A. Gautier and X. Wauthy. Teaching versus research: A multi-tasking approach to multi-department universities. European Economic Review, 51(2):273-295, 2007.

[16] A. Modirkhorasani and P. Hoseinpour. Decentralized exam timetabling: A solution for conducting exams during pandemics. Socio-Economic Planning Sciences, 92:101802, 2024.

[17] L. Shaban, E. O'Flynn, W. Mulwafu, E. Borgstein, A. Bekele, N. Bachheta, D. Stanistreet, and J. Gajewski. Factors Influencing Exam Performance of Surgical Trainees in Sub-Saharan Africa: A Retrospective Analysis of the College of Surgeons in East, Central, and Southern Africa Membership Examination. Journal of Surgical Education, 81(3):404-411, 2024.

[18] S. Moroz, R. Andrade, L. Walsh, and C.L. Richard. Student Performance on an Objective Structured Clinical Exam Delivered Both Virtually and In-Person. American Journal of Pharmaceutical Education, 87(7):100088, 2023.

[19] D.H. Murphy, J.L. Little, and E.L. Bjork. The Value of Using Tests in Education as Tools for Learning-Not Just for Assessment. Educational Psychology Review, 35:89, 2023.

[20] B. Das, M. Majumder, S. Phadikar, and A.A. Sekh. Automatic question generation and answer assessment: a survey. Research and Practice in Technology Enhanced Learning, 16:5, 2021.

[21] B.L. Moorhouse, M.A. Yeo, and Y. Wan. Generative AI tools and assessment: Guidelines of the world's top-ranking universities. Computers and Education Open, 5:100151, 2023.

[22] S. Tobler. Smart grading: A generative AI-based tool for knowledge-grounded answer evaluation in educational assessments. MethodsX, 12:102531, 2024.

[23] University of La Laguna. Notes of the subject "Operating Systems". Available at: https://ull-esit-sistemas-operativos.github.io/ssoo-apuntes/so2324/, 2024.

[24] L. Richardson. Beautiful Soup: Screen-scraping library. Available at: https://pypi.org/project/beautifulsoup4, 2024.

[25] deepset GmbH. Haystack: Open source AI framework for production. Available at: https://haystack.deepset.ai/, 2024.

[26] Python Software Foundation. Markdownify: Convert HTML to Markdown. Available at: https://pypi.org/project/markdownify/, 2024.

[27] J. White, Q. Fu, S. Hays, M. Sandborn, C. Olea, H. Gilbert, A. Elnashar, J. Spencer-Smith, and D.C. Schmidt. A prompt pattern catalog to enhance prompt engineering with ChatGPT. ArXiv preprint 2302.11382, 2023.

[28] University of La Laguna. Code repository for automatic generation of multiple-choice tests. Available at: https://github.com/ull-esit-sistemas-operativos/ssoo-mcq-gpt, 2024.

[29] Moodle Pty Ltd. Moodle: World's most popular learning management system. Available at: https://moodle.org, 2024.

 Received 28 March 2025

www.ingramcontent.com/pod-product-compliance
Lightning Source LLC
LaVergne TN
LVHW082333200926
848481LV00012B/3329
* 9 7 8 1 8 4 8 9 0 5 0 1 6 *